THE TACK ROOM

The story of saddlery and harness in 27 equine disciplines

Paula Sells

MERLIN UNWIN BOOKS

First published in Great Britain by Merlin Unwin Books Ltd, 2018

Text © Paula Sells 2018

All rights reserved, including the right to reproduce this book or portions thereof in any form or by any means, electronic or mechanical, including photocopying, recording, or by any information storage and retrieval system, without permission in writing from the publisher:

Merlin Unwin Books Ltd, Palmers House,
7 Corve Street, Ludlow
Shropshire SY8 1DB

www.merlinunwin.co.uk

The author asserts her moral right to be identified with this work
The author is donating 30% of her royalty income on all sales of this edition to the charity World Horse Welfare

ISBN 978-1-91072377-7
Typeset in 11 point Minion Pro by Merlin Unwin Books
Printed by 1010 Printing International Ltd

Photo credits

Adrian Booth 165 (bottom left)
Annabelle Perez xi
Annie Rose, Cumbrian Heavy Horses 154, 155, 158
Beamish – The Living Museum of the North 164 (top)
Blair Castle, Internet image 220
Bob Powell Archive 114
Both photos © The Tschiffely Literary Estate: Basha O'Reilly, Executrix 151
Bridgeman Images 95
CSRC 197
D. Russell 31 (top), 32, 33
Debbie Burton 7
Denbigh Museum 141, 142
Diagram by Cate Rae iv, 30
Diagram by Stella Havard 125
Diagrams by Terry Keegan 40
Edward Sells 169
Fiona Waterer 117
Florence Midwood xii
Gem Hall 93
Getty Images 109, 214
Giles Penfound 74
Graeme Cumming 219
Hunting map © Rosemary Coates 137
J. & E. Sedgwick & Co. 2
John Minoprio 15, 43 (top), 201, 202, 203, 205, 206, 207
Julian Portch 209, 210
Kevin Wright 35
Lionel Edwards Estate, courtesy of Felix Rosenstiel's widow and son 129
Lynne Munro 195, 198
Lynne Shore 45, 110, 111, 112, 170 (bottom), 171
Margaret Salisbury and Fiona Midwood 165 (bottom right)
Max Wenger 96, 98, 99
Michael Roberts 170 (top)
National Archives of Coalmining 163
National Leather Collection 25
National Museum of Wales 42, 43 (bottom), 164 (bottom)
Nicki Thorne 106, 107 (bottom)
Nico Morgan 23
North Somerset Council and South West Heritage Trust 2017 57
Pamla Toler Cover photo, vii, 4, 5, 9,10, 27, 34, 36, 46, 48, 49, 58-60, 62-64 (top), 73 (bottom), 75-77, 82 (top), 86, 91, 115, 116, 118, 119, 124, 131, 146, 156, 160, 161, 172 (bottom), 175, 178, 181, 184, 186, 216, 217, 218, 225, 226, 230, 231, 232, 236, 237
Paul Lunnon 176, 177
Paula Sells viii, x, 14, 16, 22, 38, 47, 50-56, 66, 67, 70, 78, 82 (bottom), 84, 87, 92 112, 116, 132, 135, 138, 140, 144, 145, 146, 148, 151, 156 (diagram), 157, 182, 184 (top) 185, 191, 194, 195, 201, 203, 204, 206, 223, 224, 225, 234, 235, 246
Persian Miniatures (internet) 167
Philip Brind 68
Photos Mark Romain 13
Queen Victoria's Saddle, Internet 25
Quentin Bertoux 24
Rex Features 73 (top)
Rex Features Ltd 12, 83, 85, 188, 224
Richard Green Gallery, London 81
Richard Stanton © 2000 1, 143
Rod Blackmore LRPS 103, 105, 107 (top), 121
Roger Harris 187
Roy Peckham 165 (top)
Sally Mitchell 235
Sarah Deptford 238
Sian Davies 228
Steffi Schaffler 148
Studio des Fleurs 214
Terry Davis 37, 39
The Munnings Art Museum © Estate of Sir Alfred Munnings, Dedham, Essex All rights reserved, DACS 2017 20
The National Stud 185
The Whitworth, University of Manchester 3
Tilhill Forestry 147
Tom Pilston 90
Vanessa Fairfax 20 (top), 31 (bottom)
Wadworth Brewery 64 (bottom)
Walsall Leather Museum 26
William Reddaway 153
www.Andrewreesphotography.co.uk 89

Contents

Introduction and Social History v
1. Leather, Tanning and Care 1
2. The Importance of Walsall 8
3. Livery Guilds and Apprenticeships 11
4. Saddles 18
5. Bridles 29
6. Harness 34
7. Lorinery – Bits, Stirrups and Trees 41
8. Historic and General Purpose Tack Rooms 48
9. Beach Donkeys 57
10. The Wadworth Brewery 61
11. Canal Horses 65
12. Household Cavalry Mounted Regiment & King's Troop Royal Horse Artillery .. 72
13. Coaching and Carriage Driving 80
14. The Circus 88
15. Classical and Modern Dressage 94
16. Endurance 102
17. Cross-country and Three-day Eventing 108
18. Farming with Heavy Horses 113
19. Harness Racing 122
20. Hunting 127
21. Horse Logging 140
22. Long Riders 150
23. Mounted Police 159
24. Pit Ponies 162
25. Polo and Polocrosse 166
26. The Pony Club 174
27. National Hunt and Flat Racing 180
28. Riding for the Disabled Association and Therapy with Horses 193
29. The Royal Mews 200
30. Show Jumping 208
31. Deer Stalking 215
32. Stunt and Film Horses 221
33. Isle of Man Tram Horses 229
34. Western Saddlery 233
Conclusion 239
Acknowledgements 241
Glossary 242
Index 243

The central importance of the tack room for many equestrian disciplines.

Introduction & Social History

Tack rooms have a life of their own, no matter how humble. That heady smell of leather and saddle soap has the power to evoke memories and to fire the imagination. Stand quietly for a few moments in a tack room, alone if possible, and absorb the communion of all the horses and ponies ever associated with that place, all the lives of the saddlers who made the tack, the people who used it and all their history. Feel the supple buttery-softness of oak-tanned leather, a year in the making, a gold standard for quality.

Leather is the most durable, practical and beautiful of all natural materials. It has partnered our relationship with horses since their domestication on the Eurasian steppes around 3,500 BC. All over the world it has been designed and decorated to proclaim the owner's status in society and pride in his horses, from antiquity to the present day.

I have spent most of my life around horses. My mother ran a riding school in north Wales until the start of World War II when her horses were requisitioned by the army and never seen by her again. But a few were saved by allocation to farm work and one, aptly named Careful, became my teacher when I was eight and she was thirty.

By the time I was nine I was riding up the mountain with friends of the same age on sharp little Welsh ponies. On a sheep-shorn slope overlooking the valley, we would take off our saddles, dump them in a pile and play wild bareback games of Cowboys and Indians. We fell off left, right and centre and had a wonderful time, the ponies enjoying it as much as we did.

Two years in Caithness followed, with the freedom of roaming on horseback over the heather moorlands and empty beaches. We moved to Cheshire where I was tamed by the Pony Club but I still thought the most boring task in the world was cleaning tack: saddle, bridle and headcollar.

As a teenager, hunting and eventing refined my attitude but in all that time I never gave much thought to my tack – how it had been made or even less, who had made it. I just put it on the pony, jumped into the saddle and went, and I only cleaned it when I had to.

Everything changed when I received a gift of an old headcollar from a friend. The leather was a dull black but good, although repaired in a couple of places, and the buckle had lost its shine long ago. This headcollar was huge and weighty with history. It was bigger and heavier than anything one could buy now, designed for industrial work. Stamped in the leather on one side were the letters, GWR, standing for the Great Western Railway. This once belonged to a Shire horse, probably standing about 18 hands high and weighing a ton. His job would have been to work the railway turntables turning the steam engine around, shunting carriages into sidings or hauling freight at stations. Who was this horse and what was he like?

He could have worn this headcollar a hundred years ago yet the workmanship in its leather was so good that my own heavy horse is comfortable in the same headcollar today.

In my tack room, the GWR headcollar hangs alongside tiny foal slips, general purpose saddles and various riding and agricultural bridles. It is a portal to another era. This gift ignited a new interest and sparked my exploration of tack rooms across the country and of the old and new saddlery and harness they hold. It opened a door into the world of the leather tanners, master saddlers and loriners who design and make our tack.

The strength of this GWR headcollar was intended to withstand the power of a heavy horse and the shocks and strains of his work in all weathers. In the same way, every use to which horses have been put has required specifically designed equipment. The rich variety of saddlery and harness found in traditional and modern tack rooms provides a narrative of our long partnership with horses in British culture.

The provenance of old tack can be investigated through the maker's name which, on harness, was generally stamped on the cart saddle or blinkers and underneath the skirt of expensive saddles. Thousands of general purpose saddles were also made decades ago by unknown saddlers and, when properly fitted, continue to be serviceable.

Long-forgotten details of old cavalry and coaching tack can be seen in period paintings and sculptures – and even in 19th century wooden gingerbread moulds; bakers specialising in biscuit-making often took wood carving lessons to make their own bespoke designs, such as horses which were bridled and harnessed. Good quality tack can last fifty to a hundred years or more, if properly treated. Champion & Wilton bespoke side saddles (1780-1962) often became heirlooms and are still used today. Old saddles can be restored with new trees made from traditional laminated birch, or from titanium and moulded plastics which can be altered with heat to fit a computer-generated model of any individual horse.

Now, when I first walk into a tack room I can feel its atmosphere and it comes to life through its past and present by looking at the tack and how it continues to evolve. I accompanied an elderly lady on a visit to her tack room, lending an arm as she had difficulty walking. We entered a small but beautifully appointed tack room with two sets of saddles and bridles and a side saddle in a waterproof cover. Even the telescopic cleaning hook for harness hanging from the ceiling still had its original leather sleeve to protect the bridle headpiece from possible iron stain. Everything was shining and ready for use yet it hadn't been used for thirty years. This was her shrine to the memories of a happy hunting past and she cherished everything in it from the boot jack to her old bowler hat. We stood in silence sharing the same deep attachment to the horse, the excitement of riding and all its accoutrements and reliving our experiences.

The traditional tack rooms found on farms and fine estates are the foundation of our busy and varied contemporary tack rooms. These historic 'saddle' or 'harness rooms', as they were formally known, were primarily for hunting, racing and agriculture and for carriage harness. Britain has many beautiful old tack rooms still in use today though they are becoming increasingly rare. Interest in equestrianism has continued to expand leading to the general modernisation of tack rooms where standard tack may sit alongside new designs for newer disciplines, such as Western riding or polocrosse. Each one is different but together they present a fascinating record of social history into the 21st century through the many types of tack that keep the equestrian world moving forwards.

Modern tack rooms may be in contemporary buildings and feature radiators, plastic-topped counters, washing machines for rugs and security alarms. Especially in competition or business yards, these are likely to be innovative and forward-looking and their tack, equipment and horse medications are at the cutting edge of current research in technology and veterinary medicine. They contrast strongly with the atmosphere and contents of the traditional tack rooms which carry the history of society's dependence on horsepower for transport, farming, commerce and warfare. Old tack rooms, with their glass-fronted cupboards over the fireplace for bridle

This fine Edwardian hunting tack room at Wigfair Hall, N. Wales, was built by Colonel Howard, CB, Equerry to King George VI and is in use today for riding and schooling young horses. Here, John Rees and Gerallt Brooks-Jones enjoy a joke and a glass of mulled wine while the hunt meets in the yard.

bits and wood-panelled walls, smell of saddle soap, Stockholm tar, hemp and hessian. Modern tack rooms have a different fragrance with a variety of leather balms and antibiotic sprays. Both though, have that unmistakable and unique smell of leather and horse which is remembered for a lifetime. Anyone who has been associated with horses will recall a particular tack room with a smile. That evocative atmosphere, the camaraderie, gossip and knowledge shared while cleaning tack, maybe a swig of sloe gin for Dutch courage, or just a mug of tea – all leave a vivid impression.

The tack room is the common link for all types of horse activities and the hub of their daily life. Not only is the tack stored and cleaned there but it is a social place where often the business of the day is conducted too. Tack rooms can be a treasure house of craftsmanship, a haven of memories with faded rosettes and photographs, or the command centre of a bustling racing yard, brewery or barracks.

A tack room may be a small garden shed, an old stone barn with original plough horse stalls or a highly organised and secure room as used by the Mounted Police. All serve their purpose and reflect the versatility of the horse in work and leisure.

This book is an account of my view through tack rooms to the craftsmanship and horsemanship involved with riding and working horses. It is an exploration of the tack found there and how it has developed to suit each equestrian activity.

Its design can also reflect our relationship over time with horses; the severe, high-port bits of the 17th and 18th century suggest an attitude of domination towards the horse while our modern, high-tech ergonomic bits show a greater degree of consideration for the horse's comfort.

Apart from a brand name, the individual Master Saddlers and Master Harness-Makers who make our tack do not always have a high profile link with the retailers, yet their work and the apprenticeship programmes are vital for future development and for maintaining the highest standards.

A packhorse bridge in Diggle, W. Yorkshire moors. The parapet is typically low so as not to catch the pack load as the horse passes. Until the canal network was established in the 1800s packhorses carried commercial goods all over the country. A packhorse can carry up to an eighth of a ton but a boat horse can pull a fifty-ton barge on a canal.

Each chapter describes a different equine discipline with its own particular demands so that the background to the various types of tack can be appreciated.

I have tried to capture the essence of each one, for example, the heart-in-mouth thrill and speed of polo, the total immersion in the natural world when deer stalking, the peace and satisfaction of ploughing and the colourful exhilaration of racing, so that the tack can be seen in context. As these different activities have become established, so the tack has evolved, sometimes over hundreds of years, to reach a point where form and function are perfectly united by the craftsmanship of saddlers and harness-makers.

The tack rooms and their contents described here are all actively in use, with one important exception: the pit ponies' tack room. The deep-mine stalls and tack rooms that were home to the pit ponies are now silent and empty monuments to the past. Patch, the last pony to work in deep mine shafts, died in Yorkshire in 2013. The pit ponies are included here as historical examples of the adaptability of the horse and its accompanying saddlery and harness.

The tack rooms of major transport depots belonging to the railway horses at Camden Lock, London and to the canal horses at the Roundhouse in Birmingham, have now vanished, though remnants of their stables are still there. Beach donkeys, famous in our coastal holiday resorts, also have a chapter because they are equids and still constitute a revered British institution which is part of our social history.

As with any specialty, there is some jargon here. Those in the know will enjoy it as a familiar language. Those who are not may conjure with beautiful archaic terms such as 'swingle tree', 'martingale' or 'shabraque' and nobody should feel excluded by it; the glossary opens a window into another world.

Horses and our social history

As technology progressed through the last two hundred years, peoples' lives changed, as did their horses' job descriptions and consequently the types of tack they used. Broadly speaking, apart from hunting and cavalry saddlery, the vast majority of tack in the 19th century was harness designed for horse-drawn vehicles or machinery. The archives of the Walsall Leather Museum remind us that in the 20th century, the demand for draught tack continued through both World Wars but declined in the 1950s when a steady increase in tack for riding began. In the 21st century, this is still flourishing and new materials now influence the manufacture and design of tack.

Progress and change are inevitable. Yet for horses, that change came with devastating speed when the petrol engine was developed and, most dramatically, motor vehicles began to replace horses on the land and the road from about 1910 onwards. Ironically, Ferrari and Lalique chose the image of a prancing horse for bonnet mascots, a nostalgic and fitting tribute perhaps. The pony and trap used for going to the shops and to church in the countryside, the cob delivering milk, the ploughing teams, and the remaining pit ponies were part of everyday life and are now beginning to slip beyond living memory. Of the enormous amount of horse traffic on the 18th and 19th century roads, few reminders remain – a mounting block and hitching post by a pub, the grooves where the grit in the canal horses' ropes etched the stanchions of a bridge or a rare drinking trough in a town centre.

Horses were essential for the building and trade of all our towns and cities, as well as for agriculture and mining to provide the nation's food and power.

Reminders of our dependence on draught horses can cause our hearts to skip a beat when they catch us unawares. On a recent visit to the Hoylake Lifeboat Station, Wirral, to view the new lifeboat, I was surprised to see a bridle bracket alongside the peg I was about to hang my coat on. The explanation given to me by the fifth-generation coxswain, Dave Dodd, revealed the long-forgotten but vital role played by the local draught horses in the community. Of the working horses around the town, whose everyday job it was to plough or pull carts for their owners, eight were needed to launch the lifeboat, until 1921 when a tractor took over the task.

The coxswain would fire the maroon from the lifeboat station to alert the crew and local horsemen that a ship was in distress. Ploughs were immediately unhitched, shafts dropped, a lad would jump up onto the cart saddle and, with shouting to clear the road ahead, horses would be galloped in their harness to the assembly point.

In December 1914, a grey mare in her field who had attended the lifeboat for many years, heard the maroon and raced off by herself to the station. On arrival she had a heart attack and died. Her owner, Jesse Bird, received £45 in compensation for her. Surprisingly, the horses earned more than the twelve brave oarsmen who manned the lifeboat and risked their lives. The station logbook of 23rd March, 1898, records that when the lifeboat was launched, the crew were paid fifteen shillings each and the horses' owners twenty shillings each. The skill needed to handle a team of eight heavy horses to haul the boat out at night and launch it in a storm, let alone reverse its trolley into the waves to retrieve the boat when it returned, are hard to imagine now.

The premium of speed has always driven change; horses superseded oxen, the internal combustion engine superseded horses, just as canal barges superseded packhorses and the railways superseded the canal barges. Within the last two

A team of eight heavy horses was needed to pull the lifeboat, which weighed nearly four tons, across three miles of sand to a coast notorious for wrecks.

A granite horse trough and milestone (1882) on the A541 in Caerwys, Flintshire. Water troughs were the 'filling stations' of their day and served some 50,000 horses a day in London alone in the 1880s. Most were erected by the RSPCA, the Metropolitan Drinking Fountain Cattle Trough Association (1867) and the Temperance Movement (1830).

generations some 500 years of continuity in farming with horses has been broken and an entire set of skills, their vocabulary and tools have been lost. This is reflected in the contents of tack rooms, for example, the harness for a pony and trap is now a rarity, except on car-free Sark in the Channel Islands. Horses used to be ridden many miles to meets and events or they travelled by train, at least until Dr Beeching closed 2500 stations in 1963. The war-poet Siegfried Sassoon travelled to meets with his hunter by train, the army routinely transported their

A Grade II listed sandstone mounting block, dating from the 17th century or earlier, in Nantwich, Cheshire. It is in its original place near a pub and now, incongruously, sits next to heavy traffic on the B5341.

'Waiting' by Judy Boyt at the unveiling on May Day, 2010. This life-size bronze was commissioned by the Liverpool Retired Carters' Association to commemorate all the carters and their horses who had worked in the port. The harness shown is used for horses working between shafts. For heavy loads, two horses, one behind the other (tandem) were typically seen in Liverpool. This horse is so lifelike it is easy to imagine it breathing.

One of the Kelpies by Andy Scott, in Falkirk, Scotland.

horses between remount centres and my mother and her contemporaries travelled by train with ponies to compete in shows all over the country. The introduction of horseboxes, wagons and trailers has replaced rail transport for all livestock now and aeroplanes regularly fly horses across the world. Happily, some groups, such as the Horseboating Society and the British Horse Loggers association, preserve the traditional roles of the working horse and some of the old skills of horsemanship are being re-learnt by a new generation.

The financial necessity of admitting the public to stately homes to raise money towards upkeep has caused the demise of many tack rooms. Some have escaped conversion to shops and cafés because the stable block is too far from the main house for this to be practicable, as at Arlington Court in Devon, now owned by the National Trust.

The National Carriage Collection is housed in the stable yard there and fortunately, an excellent and active tack room survives. Many tack rooms found at inns that once catered for horses working on the canals and roads are now garages. Country estate and farm tack rooms risk being converted to tearooms and workshops. A clue to a conversion can be the rounded brickwork on either side of a doorway entrance – built to prevent the horse from knocking its hips as it left the stable – which is occasionally retained by sensitive architects as a piece of our heritage.

Very few public memorials pay tribute to our debt to horses and their significance to the economy but three which have been established since the millennium are particularly poignant. A 200-metre long, coal-shale landscape sculpture of a pit pony by Mike Petts, in Caerphilly, south Wales, 2000, commemorates the millions of horses which worked underground. A stone mural by David Blackhouse, in Park Lane, London, 2004, is a memorial to all the animals, including horses, that died in 20th-century wars. The life-sized bronze of a draught horse in its 'Liverpool Gear' (harness typical of Liverpool) by Judy Boyt, in Albert Dock, Liverpool, 2010, is a tribute to the horses which had worked in the port since the end of the 18th century and particularly during both World Wars. During the World War II air raids, the carters would stand with their horses rather than abandon them when the siren sounded.

Despite the lack of horses in the urban environment today, the horse remains a significant inspiration for contemporary artists. Nic Fiddian-Green's huge bronze head of a horse drinking near Marble Arch, London, was inspired by the Horse of Selene, a stone carving of a harnessed carriage horse from the Parthenon, and provides an eloquent bridge from ancient to modern times. The massive equine steel structures in Falkirk, The Kelpies, by Andy Scott (2014) were inspired by the canal horses that used to work the barges on the Forth and Clyde canals. As a link to equine history thousands of years ago, the luxury goods manufacturer, Hermès of Paris, decorated a collection of porcelain, Cheval d'Orient, with Persian miniature paintings from the Shahnameh epic poems from the 7th century. Always innovative, Hermès, who began making harness and saddlery for the carriage trade in Paris in 1837, now makes high-tech saddles designed for different disciplines using carbon fibre, injected thermoplastics and titanium.

Working horses have not been entirely overlooked in today's fast-moving technological world and a few hundred in the UK continue to do public service in the Army, Mounted Police and in transport. Wadworth's beer is still delivered daily by Shire horses in Devizes and tram horses in the Isle of Man often give at least fifteen years of public service for the Douglas Bay Tram Horse Company. Others work in agriculture, forestry and for the canal leisure industry. The British Horse Loggers association has established new apprenticeships leading to full-time forestry work with horses. The Windsor Greys and Cleveland Bays at the Royal Mews are regularly used for ceremonial occasions and Thomas, a black horse serving in the Household Cavalry at Hyde Park Barracks, has recently retired after nineteen years' service.

The welfare of horses, ponies and donkeys is monitored by the RSPCA, World Horse Welfare and various organisations. The British Horse Society, the Pony Club and the Rehabilitation of Racehorses promote higher standards of care which are a huge improvement on the past. New approaches to communicating with horses have been initiated by Monty Roberts, Mark Rashid and others and psychiatrists are aware of the value of using horses in therapies for those with behavioural difficulties. Our relationship with the horse continues to develop in the 21st century and to be mutually beneficial.

Still Water, Marble Arch by Nic Fiddian-Green. The busy London traffic contrasts with the peace and quiet of a horse drinking.

CHAPTER ONE

Leather, Tanning and Care

Leather is one of the most remarkable natural materials. Its strength, suppleness and durability make it an ideal choice for harness and saddlery.

The discoveries of a Scythian saddle sealed in an ice-filled Siberian tomb about 500 years BC, a 400-year-old, leather-covered saddle tree excavated from black mud in Southwark and leather mittens recovered from the Mary Rose, Henry VIII's warship which sank in 1545, are all examples of the power of tanning to help preserve leather.

Almost all of the leather made into harness and saddlery now comes from cattle hides. The size of the hide and the quality of the tanning are important; it should be large enough to make a matched pair of 60-inch long reins (from neck to tail, either side of the spine) while the quality of tanning determines the suppleness and strength of the leather. Different thicknesses are required depending on the use, for example, show bridles (3.5mm) and for stirrup leathers on a polo saddle (5mm). Full grain (best quality) leather retains its natural strength and breathability while top-grain or split leathers are less durable.

Vegetable tanning, using the tannins from tree bark, is a lengthy process taking at least a year and produces the strongest leather; it has a breaking strain of approximately four tons per square inch. Mineral tanned leather is processed with chromium salts or alum and is quicker and cheaper to produce (about two weeks). It is very supple but not so durable or strong and is usually used for shoes, clothing and furniture. Smoking leather, particularly buckskin (deer skin), is another method of preservation which was widely used by Native Americans, though for clothing rather than saddlery.

Leather from goats – Morocco leather, famous for its beauty and colour since the 16th century – was used for the royal harness and sheep leather (see Pittards plc, page 5) is used for accessories such as riding gloves. Japanned leather, also known as patent leather, is now plasticised rather than shellac-varnished.

Pigskin is very hardwearing and was used on hunting saddle seats but is more difficult to obtain now, while doeskin with its suede finish, is popular for equitation and side saddles. The most expensive leather is rose-tanned where 'otto of roses' (oil distilled from the petals of roses picked before sunrise) replaces the usual nourishing oils and the fragrance can last for years.

The finest leather known is believed to have been Russian leather which had exceptional durability and resistance to water. Willow and birch bark were used for tanning the leather which was then treated with birch oil. It was dyed a rich red colour using sandalwood and was apparently greatly sought after by Marco Polo and the Tartars. The distinctive aroma of birch oil became synonymous with luxury and it was used as a fragrance in products by Coco Chanel and for Cussons' Imperial Leather soap. In 1786, the brigantine *Metta Catharina*, carrying a cargo of Russian leather from St. Petersburg to Genoa, was wrecked and sank in the English Channel. In 1973 the cargo was salvaged from the disintegrated hull and the leather found to be intact. After cleaning and nourishment a few of these hides were purchased by Hermès and made into handbags and briefcases. To this day, the exact tanning procedure for Russian leather is unknown despite attempts to reproduce it.

'Cuir bouilli' is leather which has been boiled to harden it and was used for boots worn by coaching postilions. Iron bands also strengthened the boot to protect the leg from being crushed by the wooden pole or the horse alongside.

Tanning at J. & F.J. Baker & Co. Ltd, Colyton, Devon

Here is leather fit for royalty. The celebrated company of J. & F.J. Baker supplies the leather for Her Majesty The Queen's carriage hoods as well as for the boots and bridles of the Household Cavalry. Finest quality full grain leather is prepared in the traditional way by tanning with oak bark dissolved in water. Sole leather for high quality shoes is also produced here, some of which sell for over a thousand pounds a pair in Europe.

In medieval times most large towns situated near rivers had tanneries. The dreadful smells from these places were due to the rotting carcases and ordure used in the tanning process. It is said that stray dogs were rounded up and kept in a cage above a pit in which the cow hides were laid. The dogs were well fed on the flesh scraps and their excrement dropped through the cage floor into the tanning pit. Pigeon and hen droppings were also used to 'bate' the skins (the enzymes they contained helped to remove the lime in the tanning process). Fortunately, improvements in technique ensure that tanneries no longer advertise their presence in the air; J. & F.J. Baker's premises are unobtrusively tucked into the grey stone walls of Colyton in Devon and there are no complaints from the neighbours.

Tanning has taken place on the banks of the River Coly since at least as far back as Roman times when Axmouth was an important Channel port. The combination of soft water, good grazing

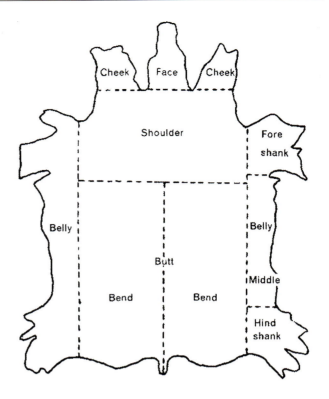

Diagram of a hide.

for local cattle and the commitment of the Baker and Parr families who have run the tannery for six generations provides the essential conditions. J. & F.J. Baker's original and extensive buildings are the most traditional of the few vegetable tanneries left in the UK. A huge Victorian waterwheel powers a machine for chopping oak bark from Wales and the Lake District. Some eight to twelve tons per year are used and the chips are soaked to produce tannic acid liquor. Various supplements can be added such as valonia (acorn cups from Turkey), acacia and sumac which affect the final colour of the leather.

Vegetable tanning is very slow (fifteen months) and expensive. The advantage of using vegetable tannins to cure the hides is that they produce superior tensile strength and durability of the leather which is important for heavy-duty use, such as for agriculture. The point of tanning leather is to prevent putrefaction and to preserve its strength and suppleness. Demand in the past was chiefly for horse harness and saddlery for the cavalry, pit ponies, agriculture and carriage driving. The importation of French oak bark for the tanneries ceased during the Napoleonic Wars and the high demand for bark drove up prices so much that special laws were enacted to protect our native trees from being de-barked.

The journey of a hide from abattoir to bridle butt takes fifteen months. Hides from Hereford cattle which are at least two years old are preferred (Friesians have bony hips which mark the hide) as these are long enough from neck to tail to provide the length of a pair of reins. The numerous stages in the pre-tanning, tanning and finishing (setting, staining and dressing) are done almost entirely by hand. The tanning process itself takes twelve months, after which the hides may be termed leather. Finishing them with oils and fats takes a further three months. Despite the nature of the business there were no unpleasant smells in the tannery, the strongest coming from a warm bucket of dubbin (mutton tallow and fish oils) which is rubbed in by hand as a final finish.

Briefly, the method of tanning is as follows: the cattle hides are soaked in a pit of hydrated lime solution (pH13) and then de-fleshed and de-haired by machine. This causes the collagen fibres to swell and readily accept the tannins. The hides are then thoroughly washed to lower the pH, squeezed and scraped ready for tanning. Each is individually folded over ropes and lowered into deep oak-lined pits of tannin liquor. The tanning shed is a large building with a floor consisting entirely of pits about six feet deep separated by narrow walkways. There is no heating and the dim light reflects off the crust of mould formed on the liquor surface.

Bark peeling. Watercolour by Samuel Lucas, 1805–1870. Bark, especially that of oak, was once a valuable commodity. Its price reached a peak during the Napoleonic Wars when it had to be home-produced.

Dehairing. Hides about to be lowered into a lime bath before moving to the tanning stage in oak bark liquor.

Over a period of three months the hides are moved by hand through a series of pits beginning with the weakest tannin solution and finishing in the strongest concentration. Here they lie, layered flat and undisturbed for a further nine months absorbing tannic acid. The collagen fibres in the hide become cross-linked and irreversibly preserved in a strong matrix. The tanned leather emerges stiff and soaked and over the next three months it is dried and the suppleness and waterproofing restored by rubbing in particular oils and fats, depending on the desired product. This final finishing including dyeing, dressing, embossing, etc. of the leather is known as currying.

English leather has been an international byword for quality for hundreds of years. It is still a craft where man's skill is superior to machines and where the eye, hand and experience create the quality of leather which will last for generations.

Currying at J. & E. Sedgwick & Co. Ltd, Walsall

J. & E. Sedgwick & Co. Ltd., renowned for the superb quality of their leather, are the last remaining curriers in Walsall. Fine leather with the necessary tensile strength for saddles and bridles accounts for much of Sedgwick's output. Every bridle strap made from Sedgwick's leather should be able to pass the 'mandrel' test: the strap should bend round a mandrel (a cylindrical rod) or pen without resisting or cracking – a test many imported leathers would fail.

The history of the industry is evident from a walk through the factory. Beginning in the comfortable, spacious office of the Director, Richard Farrow, and under the watchful eye of his black Labrador who comes with him to work every day, is a large watercolour painting of bark-peeling (by Bates, 1895) in Cannock Chase, not far from Walsall. The bark was used to provide the tanning liquor for at least six local tanneries at that time. On the office wall is a restored measuring device made of wood and iron which, in Richard Farrow's apprentice days, was the only way of measuring the area of a hide. The metal arm which was attached to a graduated wheel was guided by hand round the irregular outline of the hide but now measurement is computerised.

Further on in the processing sheds is a machine the size of a car for splitting hides into variable thicknesses. Less than fifty years ago, splitting and shaving were skilled manual jobs and an expert could complete twelve hides an hour. This computerised splitter is accurate to 0.2mm and can process hundreds of hides every hour.

Beyond the splitting machine are the sheds where the currying process on the vegetable tanned hide begins. This involves many steps in a three-month-long process but briefly, a short re-tannage of the hide, including dyeing in large wooden drums to correct any unevenness, is followed by natural drying, setting and lubrication. Along the entire length of the shed are massive benches of Welsh slate which provide a flawless surface on which the curriers smooth out the leather and lay the fibres on the flesh side ('glassing').

The leather is rubbed with mutton or beef fat and cod oil (a ton per month is used in the factory) to replace losses during the tanning process, dyed to the desired colour by hand and finally the grain sealed with a mixture of beeswax and oils. It is easy to see why the phrase 'elbow grease' originated in this industry as the curriers at the bench bend over the hides applying their weight to the task. 'Time',

Leather drying naturally without artificial heat. Note slatted floors for air circulation.

emphasises Richard, 'is the most important ingredient of the entire tanning and currying process. Cutting corners and hurrying always results in inferior leather.' At least three quality control checks are carried out at different stages in the processing and the slightest imperfection will consign the hide to the bin. Richard is the third generation of his family to own Sedgwicks and has combined the modernisation of the factory with traditional methods to produce an output of some five hundred top quality hides per week for a thriving global market.

Following his recent retirement, Sedgwick & Co. is now under new ownership.

Innovation in tanning at Pittards plc, Yeovil, Somerset

Pittards in Yeovil has been tanning and finishing leather on the banks of the River Yeo since 1826. The company has developed many of its technical leathers in conjunction with British event riders, including Ian Stark. Ian, who has won the Badminton Horse Trials three times, holds four Olympic medals, a European gold and numerous other awards, has used a Pittards' helmet, gloves and saddle in competitions. He is now a course designer, coach and commentator. Ian was renowned for his verve and skill round cross-country courses, usually on forward-going horses, such as Murphy Himself who pulled like a train. Good gloves are essential for gripping wet or sweaty reins at speed and the highly engineered leathers produced by Pittards are designed for extreme performance.

Surprisingly, most of the leather used for these gloves comes from Hair Sheep in Ethiopia and is known as Cabretta leather. This breed produces skins with fine hair follicles and a high tensile strength and much of it is prepared in the Company's local tannery and manufacturing facility there, an important source of employment for the Ethiopian community.

Pittards has pioneered innovative technology which has greatly expanded the versatility of leather making it waterproof, washable, antimicrobial, flame retardant, stain and abrasion resistant but still breathable and supple.

The skins arrive as 'wet blues' (referring to the colour resulting from chromium sulphate tanning) and are ready for the re-tannage in which polymers and micro-ceramic particles are introduced. These help to prevent the glove leather from deforming or stiffening in rain or sea water or when soaked in perspiration yet enable it to have a strong grip on slippery surfaces. Most of the processing is computerised but every skin has to pass through inspection by hand after each stage, a skill which the apprentices here will develop over time.

Since providing the World War II Spitfire pilots with gloves Pittards now have an international market supplying the armed forces, round-the-world yachtsmen, mountaineers and top international sportsmen as well as a large retail store full of luxurious clothing and accessories.

Maintenance and care of leather

Leather, rather like wood, retains elements of its living origins. The secret of maintaining its suppleness and strength is to treat it as naturally as possible. After use, saddlery and harness should be sponged clean and hung up to air dry away from artificial heat sources. When dry, glycerine soap applied with a barely damp sponge is simple and effective and good products containing beeswax, fish oil and citronella are pleasant to use and may help to deter flies in the summer.

The structure of leather fibres has been altered by tanning to prevent putrefaction and, in the process, its natural lubrication has been depleted. To compensate for this, oils, fats or waxes must be rubbed in at regular intervals. It is important that the origins of the oils and fats are either animal or plant derived. Mineral oil, often added to dilute natural oils in commercial products, is particularly harmful to leather. Over-oiling can be as damaging as leaving the leather to dry out. The mineral oils act as a 'driver', pulling more oil deep into the structure of the leather which will eventually rot. Because it is rapidly absorbed, more oil is often applied and in extreme cases the leather becomes soggy and structurally unsound. Stitching can also rot, especially if the thread is linen, though polyester thread is commonly used now which is more durable.

No artificial heat or glycerine sprays containing alcohol should be applied as both will dry out the surface. The most common mistake is to place wet saddles and bridles close to a radiator where the accelerated drying will quickly cause damage. Another habit which is bad for leather is to rinse off the bridle bit in a bucket of water after use without drying it. This will eventually weaken the leather (billet) next to the bit unless the bridle is regularly taken apart for cleaning and nourishing the leather.

Metal polish applied to rings, buckles, stirrups or harness terrets is easily accidentally spread onto the turns of leather to which they are attached. The junctions between leather and metal are always the first to wear and if the leather at this point is also damaged by the corrosive polish, it will be the first part to snap under stress.

Mice are attracted to the oils and fats used on the leather and unless there is a cat on duty in the tack room they can cause damage to leather, saddle linings and fabric.

An Ethiopian Hair Sheep from which high quality glove leather is produced.

Holy Saddles

Mould on leather does little harm apart from looking unsightly and can be discouraged by the use of a dehumidifier in the tack room. Leather mould may, in fact, be beneficial as this Irish anecdote suggests.

In the central Athlone region and probably beyond, there is the well-known phenomenon of 'the holy saddle'. If a person should have a cut on an arm or leg, the local priest can be asked to bring a 'holy saddle'. After a few prayers, the patient keeps the saddle as close to the wound as possible for two or three days. The priest returns, removes the saddle and very often finds that the cut has completely healed.

This may be due to divine intervention or the placebo effect but there is an alternative possibility. In the damp Irish climate, leather in storage quickly becomes mouldy. It is known that the mould used in the making of Dorset Blue Vinney cheese reputably came from a harness mould and that it is of a penicillin type. The mould on a 'holy saddle' may therefore contain enough penicillin to expedite the healing process.

The use of a variety of moulds for treating wounds and infections has been known since 1500 BC. A number of English physicians had also noted their antibacterial effect before the French medical student, Ernest Duchesne, succeeded in making extracts from moulds which successfully treated experimental infections in guinea pigs in 1897. Duchesne had apparently got the idea from Arab stable hands who traditionally used saddle moulds to treat wounds and infections.

In 1925 Alexander Fleming was credited with having 'discovered' penicillin which was later developed into a commercial medicine by Chain, Florey, Jennings and many others in 1942.

CHAPTER TWO

The Importance of Walsall

Before the town of Walsall in the West Midlands became enveloped in Greater Birmingham, it was the world manufacturing centre of equestrian goods for nearly two hundred years. Just about everything a horse and its handler might need was produced here: saddlery, harness, grooming brushes, decorative brasses, rugs, whips, gloves, tools, stable fittings etc. and elaborate catalogues were sent to all corners of the British Empire.

Thomas Newton of Walsall, a third generation loriner, is credited with the first successful expansion into saddlery-making about 1830. He made bespoke harness for the Empress of Austria and for Indian maharajahs and produced an extraordinary and beautifully illustrated work *The Saddlery of All Nations* about 1872, the first edition of which is in the British Museum.

The British Army was the single biggest customer with cavalry requirements for the Crimean, Peninsula and Boer wars as well as both World Wars. More than a hundred thousand saddles were produced in just the first two years of the First World War by D. Mason & Sons (who also produced the dog harness for Captain Scott's expedition to the South Pole). The nickname for Walsall Football Club is still 'The Saddlers'.

The Romans had recognised that Walsall was rich in natural resources and used the water supply which flowed from Wales through limestone for their tanneries. By the 15th century Walsall had developed as a metal working centre due to local supplies of wood and coal, iron ore and limestone which was hauled out of the mines by pit ponies. The limestone was used as flux to smelt iron ore for the lorinery industry. Loriners are specialist metal workers who make all the metallic components of saddlery and harness; bits, buckles, saddle trees, etc., though not horseshoes. In these early times many English surnames were acquired through the trade such as Skinner, Tanner, Barker and Sadler.

Walsall was surrounded by good pastureland for the grazing of local cattle herds which provided leather and water from the River Tame supplied the factories and tanneries. Conveniently centrally located in the country for canal, road and rail communications, the equestrian industry flourished.

It is estimated that Victorian everyday life depended on some three million horses for transport and agriculture as well as on a huge pit pony population. The equestrian leather industry peaked in Edwardian times when over a hundred saddlery and harness companies were recorded here in 1900. This was not to last though and within a decade the arrival of the motor car began the decline of the equestrian trade.

After World War II, diversification into fancy goods for export was successful until cheap imports undercut prices in the 1960s. Competition today in the equestrian goods market from imports of poorly tanned leather, particularly from India and China, together with the popularity of nylon webbing, has reduced the number of leather manufacturers to about sixty companies connected with the manufacture of saddlery. These companies specialise

One of the Albion Saddlery workshops, Walsall, West Midlands.

in both traditional and innovative top quality goods and include several Royal Warrant holders. 'Made in Walsall' has always been a guarantee of quality and while the volume of equestrian goods manufactured there exceeded any other centre, there were also a number of distinguished saddlers in London, for example, Champion & Wilton, Owen, Mayhew until 1940, and Gidden and in Newmarket, F.E.Gibson.

One of the innovative saddle makers in Walsall is Albion whose clients include several Olympic teams and riders of various nationalities. Established in 1985 by Paul and Sherry Belton, both trained in classical dressage by instructors from the Spanish Riding School, it is thriving as a family firm with their children taking an active part in its daily running. Paul Belton, who also competed at Wembley and Hickstead, is a past president of the Society of Master Saddlers and a qualified engineer. He combines all these skills to design a range of highly technical saddlery, often working with riders such as William Fox-Pitt and Laura Tomlinson to test out new ideas.

The conformation of horses today is very different from those ridden by our parents' generation; the warmblood continental breeds are bigger and rounder than the narrow Irish/thoroughbred hunter type and horses now generally tend to be fatter. Styles of riding have also changed, for example in eventing, polocrosse and show jumping; speed and agility require robust and supportive saddles. The old general purpose saddles are no longer good enough for the high standard of competition today. Now we have the technology to precisely measure and reproduce the shape of a horse's back and from it make an exact mould from which to build the saddle. Computerised pressure pads, laid between the saddle and the horse's back, are among the techniques available for manufacturing a variety of metal, wood and synthetic trees which fit the individual horse more closely and comfortably than ever before. Contact between the saddle and horse can be analysed while the horse is moving whereas most fittings were previously done while the horse and rider stood still.

Laminated birch saddle trees being trimmed in the Albion workshop. Birch provides strength with flexibility and is sourced from Scandinavia where the growth rate of the tree provides the correct density and tensile strength. Six veneers of 1mm birch are glued and placed over a mould in a heated vacuum press for up to three hours while the glue is activated by infra-red light. The resulting laminate is scrimmed and varnished to retain the wood's natural moisture.

In 2004 Albion produced an adjustable tree which allowed alteration of the head (pommel), seat area and cantle to be made at any time during the saddle's life. This can be done as well as, or instead of, reflocking – depending on the condition of the horse and the degree of saddle wear. This principle of an individual saddle which can adjust to the seasonal changes of shape of a horse is under continuous development in the world of high-tech master saddlers. Saddles are also flexibly designed to suit the rider with variations in flap lengths and supportive padding round the leg position with optional extra panels or inserts.

The flocking used (wool, memory foam or air) and the type and position of the girth and stirrup bar fittings are all-important components of a saddle which suits both horse and rider.

The Society of Master Saddlers runs courses for fitting saddles and a qualified fitter can confirm that the saddle or bridle is correctly constructed to suit each individual.

Quality harness and saddlery are expensive to buy but are long-lived and hardwearing, particularly for agricultural, carriage driving, polo and hunting use. Saddlery for leisure riding which is more subject to fashion is not required to be as robust.

The City of London livery companies such as the Worshipful Companies of Saddlers and Loriners sponsor apprenticeships and placements to maintain a manufacturing skills base for the equestrian world and to monitor quality.

CHAPTER THREE

Livery Guilds and Apprenticeships

The City of London is the centre for most of the craft guilds or livery companies which originated in Britain in the Middle Ages and many still have their headquarters in a Hall in the City today.

Guild organisations are found throughout Europe. They were formed as trade associations to support and regulate trade and craft. Their objectives were the maintenance of manufacturing standards, arbitration in disputes, welfare of members and apprentices, education and charitable donations and they continue to be relevant today. Companies or Fellowships are thought to have existed here before 1066 and are also called guilds and sometimes 'mysteries', a term deriving from the Latin 'misterium' meaning professional skill. Each guild usually had a coat of arms, a patron saint, a link with a particular church and its own distinctive clothing worn by qualified members, hence the name 'livery' company.

The ubiquity of the horse in every aspect of civic and rural life over the centuries led to the formation of several livery companies concerned with supplying equestrian goods.

Today the Worshipful Companies of Saddlers, Cordwainers (leather workers), Loriners (metal workers), and Coachmakers and Coach Harness Makers play a significant part, together with other organisations, in influencing the design of saddlery and harness through research and teaching. The Farriers' and Curriers' (leather finishers) livery companies are also of central importance to those connected with horses.

In the past, the manufacturing priorities were mainly quality, durability and cost. With the introduction of new analytical methods for measuring the horse's health and physical interaction with its saddlery, the comfort of the horse is an increasingly major consideration. As with their predecessors, the next generation of saddlers will have been taught, supported and examined by the livery companies in the light of contemporary knowledge.

The City of London livery guilds take part in the Lord Mayor's Show on the second Saturday of every November. The Lord Mayor's coach was designed in 1757 by the architect Sir Robert Taylor and is pulled by six heavy horses. The painted panels by Cipriani show the City's guardian spirit, Genius, receiving elephants' tusks, an Arab horse and a lion which represent the City's trading. Its moral qualities of truth, temperance, justice and fortitude are also represented as is the Past and Future by views of St Paul's dome before and after the Great Fire.

The Worshipful Company of Saddlers (1363)

The Lord Mayor's coach and six heavy horses, with guild members in full livery, parade through the streets as they have done for the last 478 years.

The Saddlers' Company motto 'Hold Fast and Sit Sure' holds a mortally important meaning. The saddle design was vital to survival in fighting and jousting in medieval times. Its high pommel and cantle helped to stabilise the rider and to protect his vital organs from the lance. Once a knight in armour had been knocked off his horse, escape was impossible and he was vulnerable to attack.

The present Saddlers' Hall in the City of London is a magnificent successor to previous halls and has been rebuilt three times. The first was ruined in the Great Fire of London and the second burnt down when neighbouring premises caught fire. During an incendiary raid on 29th December, 1940 the Hall was bombed and destroyed for a third time. The tide was out and the fire could not be extinguished for lack of water. Much was lost but the Company's wine cellar survived. Today, the Company has many fine treasures on display including the side saddle cover thought to have been used by Queen Elizabeth I in 1574 when she visited Bristol.

The relationship between the Saddlers and Royalty in earlier times caused a waxing and waning of the company's assets. Whenever a war had to be financed, the monarchy would demand a handsome contribution (usually provided by melting down the Company's silverware). During the Civil War in 1643 the Company had to sell all its silver plate to support the parliamentary party against Charles I. However, in the days of cavalry warfare, riches were recouped through substantial orders from the Crown for saddlery and harness.

In the 21st century, tradition and progressiveness meet effectively in the Company's policies; the highest standards in manufacturing are maintained, research and teaching are a high priority for their charitable support and they work closely on issues affecting the smooth running of the City. A wide range of equestrian organisations benefit from their support, such as British Eventing (*see* page 108), Riding for the Disabled (*see* page 193), the King's Troop (*see* page 72), and the Household Cavalry (*see* page 72).

The Saddlers' Company also provides substantial funds for the training of young saddlers, such as the Apprenticing Charity (originating from a bequest from Richard Banner in 1698) as well as setting the standards for qualification and examination.

Apprenticeships

The Saddlery Training Centre founded in 2000 is a modern unit on the outskirts of Salisbury and is closely associated with The Worshipful Company of Saddlers, The Society of Master Saddlers and a number of organisations involved in teaching and the equestrian trade. Its founder and director is Master Saddler Mark Romain, MBE, who, together with Master Saddler Richard Godden (side saddle manufacture and restoration) are the course instructors.

Mark Romain, MBE, Master Saddler and Director of the Saddlery Training Centre, Salisbury, surrounded by students on the Millennium Apprenticeship scheme supported by the Worshipful Company of Saddlers.

Danish student Dan Pedersen, working on the panel of his General Purpose saddle. In addition to his daily work as a saddler, he attends an annual six week course for three years at the Saddlery Training Centre to gain his apprenticeship with the Worshipful Company of Saddlers.

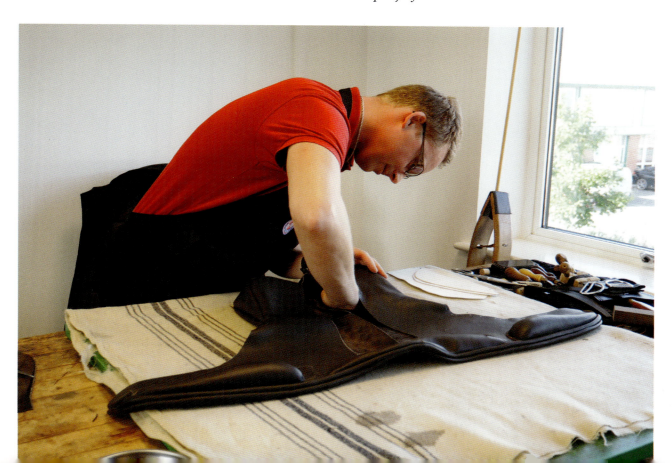

The courses here cater for students who have already had some experience of working in saddlery, such as holding a Princess Royal College (Capel Manor) two-year diploma, and who wish to become qualified saddlers. They may then establish their own businesses independently and contribute to the rural economy or work in the saddlery industry.

Traditional methods for handmade saddlery and harness-making are taught throughout with the emphasis on handstitching, though advanced students receive tuition on the use of the latest industrial sewing machines. The pieces made for skill assessments are of British leather and both saddle trees and stirrup bars must conform to the British Standard BS 6635 2015, introduced in 1999. The highest quality of workmanship for both the welfare of the horse and the safety of the rider is fundamental for all apprentices.

As Mark Romain explains, 'The technology is moving fast with the introduction of new synthetic materials but by understanding the principles of construction and fitting, these apprentices should be able to adapt their work to any innovations in the future.'

Apprenticeship bursaries, which cover some or most expenses, are awarded on a competitive basis, as there are more applicants than places. They are funded by a variety of organisations such as the Worshipful Company of Saddlers and/or the Skills Funding Agency, depending on the student's background.

Students who are awarded an apprenticeship will already be employed by a Master Saddler or Qualified Saddler.

In all cases, it is usual for a minimum of four years to be spent working both at their place of employment and also attending courses or modules for about six weeks a year at the Saddlery Training Centre. Apprentices who have qualified after four years may wish to return for further specialised courses that can lead to membership of the Society of Master Saddlers (formed in 1966).

The National Lorinery Collection of William Albery in Horsham Museum, Sussex.

LIVERY GUILDS & APPRENTICESHIPS

During their first four years of training they will have attended saddle and harness-making and fitting courses and achieved City & Guilds qualifications. Their work is regularly assessed and pieces they have made may be entered for the National Saddlery Competition, judged by the Society of Master Saddlers and hosted by The Worshipful Company of Saddlers at Saddlers' Hall in the City of London.

I was privileged to be invited to the annual National Saddlery Competition and prize giving ceremony. It was a grand occasion in the Great Hall which was filled to capacity. The majority of the apprentices were women, perhaps reflecting the bias among those who ride or work with horses. The first lady Master of the Company, Petronella Jameson, was elected in 2013 and HRH The Princess Royal is now the Perpetual Master. Judging by the outstanding quality and innovation of the work on display, the Saddlers' Company must be proud to see their apprentices achieve standards of international excellence. Such pride is no doubt matched by the great satisfaction felt by each apprentice in having created a piece of saddlery by hand which will be used for several lifetimes.

The Worshipful Company of Loriners (1261)

Loriners design and make the metal components used by equestrians, such as stirrups, buckles, saddle trees and bits but not horseshoes. Shoeing is the province of the Worshipful Company of Farriers.

The lease on the Loriners' Hall near London Wall, mentioned in Samuel Pepys' Diary in 1668, was not renewed and the Loriners moved out of London in the 1800s. They are now based at The Princess Royal College of Animal Management and Saddlery (previously Capel Manor College) in Enfield, London, which offers a Diploma in Saddle, Harness and Bridle Making, the only full-time saddlery course in the UK. On this two-year course, students are taught traditional techniques in saddle-making and fitting, such as cutting, handstitching and braiding while the related lorinery work is demonstrated at

A 1978 Rolls-Royce Phantom VI from the Royal Mews with the King's Troop Royal Horse Artillery at Windsor. The Worshipful Company of Coachmakers and Coach Harness Makers (see overleaf) adapted their skills in the Edwardian era to building coachwork for the motor car.

Abbey England, Walsall, the major manufacturer of equestrian lorinery in Britain. There they learn about the metallurgical aspects of saddlery, for example, how metal components are fitted to leather, how steel 'springs' for a saddle tree are tempered to provide tensile strength and how to make bits as comfortable as possible for a horse's mouth. Young students from all over the world can achieve the diploma and then go on to a four-year Millennium Apprenticeship and/or a two-year Government Advanced Level 3 Apprenticeship with a recognised saddler before qualifying. These courses are supported by the several livery companies on the Advisory Board: Cordwainers, Saddlers, Loriners, Coachmakers and Coach Harness Makers and Curriers.

About a dozen places are available every year and the course is always oversubscribed. With their transferable skills, qualified students can diversify and several commented that understanding how to work with leather gave them the knowledge and freedom to design successfully in fashion and for the manufacture of small leather goods. Jimmy Choo, of celebrity shoe fame, graduated from this course with honours in 1983.

The Loriners' company has donated funds for many equestrian projects; particularly to the King's Troop, Household Cavalry (*see* page 72), Riding for the Disabled (*see* page 193), Pony Club (*see* page 174) and veterinary research. HRH The Princess Royal is an Honorary Liveryman of the Company and was the Master in 1992.

National Lorinery Collection

The National Lorinery Collection is in Horsham Museum in Sussex. It is based on William Albery's extensive personal collection which he bequeathed to the town in 1950. William was born in Horsham in 1864 and left school aged twelve when he was apprenticed to the family's saddlery business.

An apprentice worked a 64-hour week for six or seven years in those days, earning a shilling a week, until he qualified as a journeyman saddler and collar maker. His father died when William was 21 and he supported his family and seven sisters by expanding the saddlery business, eventually becoming President of the Master Saddlers' Federation in 1935. A popular figure in the town, he was the founder of the Horsham Borough Silver Band and wrote many articles about the saddlery trade.

Billingshurst farrier Grant Laing's mobile smithy. Shoes may be handmade or adapted to each hoof from standard blanks imported from Malaysia or Holland.

The Worshipful Company of Coachmakers and Coach Harness Makers (1677)

The Company was finally given its royal charter by Charles II in 1677, its proposed amalgamation with the Wheelwrights having been abandoned when interrupted by the Civil War. According to a 17th century writer, the coachbuilding trade was a combination of carpenters who formed the carriage structure, tailors who upholstered the inside with cloth, silk and velvet and the shoemakers who made the leather hoods. The harness makers, although included in the Company, had a stronger working affiliation with the leather related Companies such as the Saddlers and Cordwainers. The painstaking varnishing and painting of the coach with at least fifteen to twenty-two coats (still practised today at the Royal Mews in London) and the coach building work of many craftsmen such as chasers (ornamental artistry), was also highly skilled.

One of the Company's objectives was to ensure that manufacturing standards were upheld. Its officials carried out spot inspections of premises and were empowered to issue fines wherever they found inferior materials, for example, old iron in a new coach, old hay and stuffing, old leather and unseasoned wood. Our English word 'coach' probably comes from the Hungarian 'Kocs' which is a town near Budapest where the first coaches are thought to have been built.

In 1900 the coachbuilding trade in Britain was flourishing but by 1910 the motor car had caused a severe decline and many coaches were broken up for scrap. In a time of abrupt change, the Company's Latin motto, 'Phoebus (the sun) Rises after the Clouds' must have been reassuring. With great foresight, the Coachmakers' Company was among the earliest of the livery companies to promote technical instruction for its members and apprentices and they adapted their skills quickly to move with the times.

Many important innovations were introduced by the Company members besides a number of advances in the design of springs and axles. Their inventions include the 'unimmersible' lifeboat

(Master Lionel Lukin, 1793), the fish-plates which join lengths of rail together (William Bridges Adams, 1837) and the first caravan. This caravan (eighteen feet long) travelled all over Britain drawn by a pair of horses, Polly Peasblossom and Captain Cornflower and was made for a naval surgeon, Dr William Gordon-Stables who became the first president of the Caravan Club in 1907.

Surviving coachbuilders diversified into building bodies for motor cars, such as Austin, Siddeley, Talbot and Daimler – the body style of early Daimlers in 1898 was taken from the Marseille Phaeton. The streamlined bodywork of Blue Bird, in which Sir Malcolm Campbell exceeded 300mph in 1935, was built by a team headed by Piercy, a panel beater from the coach builder Gurney Nutting of London.

Aviation pioneers such as Sir Frederick Handley Page and Sir Richard Fairey were admitted to the Livery in the 1920s and with the outbreak of WWI, members of the Coachmakers played a significant part in the building of aeroplanes such as the Avro Lancaster and Halifax bombers, as well as tanks, airborne radar scanners and components for Spitfires and Hurricanes.

The Company's interests continue to bridge the coaching era and contemporary manufacture. Mark Broadbent, a Company Liveryman, founded Fenix Carriages which builds and renovates all types of carriages. In 2001, Fenix Carriages were commissioned to build a Millennium Coach, modelled on an 1850s road coach and drawn by four horses (also trained by Mark and Joanna Broadbent). The coach was entirely hand-built by the three in-house craftsmen whose skills include all the iron working, spring making, painting, wheelwrighting and fabrication. Fenix is one of probably only a handful of companies worldwide able to undertake the entirety of such a project.

A fine collection of horse-drawn vehicles of the late George Mossman (Coachmaker Liveryman and recipient of the Company's Gold Medal in 1992) can be seen at Stockwood Museum in Luton.

The Coachmakers' Company recognised its earlier roots by the establishment of an apprenticeship at the Royal Mews in 2014 and continues to provide student bursaries and awards for the automotive and aerospace industries.

The Worshipful Company of Farriers (1674)

The Livery Company is the examining body for farriers and apprentices. They cover a range of regular and specialist remedial work in all equestrian disciplines. Except for bespoke shoes and for farriery competitions, imported shoe blanks are usually used and individually adjusted as necessary. Many tack rooms have basic farriery tools for removing horseshoes in the event of a problem such as an abscess or the shoe being pulled half off and being a hazard. Anyone may remove a shoe but it is illegal for an unqualified person to put on a shoe.

The importance of farriery cannot be over-estimated and is summed up in the old poem:

> *For want of a nail the shoe was lost.*
> *For want of a shoe the horse was lost.*
> *For want of a horse the rider was lost.*
> *For want of a despatch the battle was lost.*
> *For want of a battle the war was lost.*
> *For want of a victory the kingdom was conquered.*

Origins of the lucky horseshoe

The story of how the horseshoe became a lucky charm describes a meeting between St Dunstan and the Devil. St Dunstan was born in about 910 in Glastonbury, Somerset and became archbishop of Canterbury in 959. When still a monk he was a blacksmith by trade. One day he was asked to shoe the Devil's horse and he also nailed a shoe to the Devil's hoof. This caused the Devil great pain and Dunstan only agreed to remove the shoe if the Devil promised never to enter any premises where a horseshoe was hung over the door.

CHAPTER FOUR

Saddles

'Hold Fast and Sit Sure' is the motto of the Worshipful Company of Saddlers. With a saddle, and the help of stirrups, the rider has a secure position to brace himself against the strength of a horse and keep control with the reins.

From ancient times to modern day, the design of saddles has progressed to fulfil the three essential requirements: control of the horse, comfort for the horse and comfort for the rider.

Cleaning tack is not always a favourite task but it is then that one really appreciates how cleverly the design of a saddle is put together – the interlocking planes and curves which miraculously form a comfortable interface between the backside of the rider and the back of a horse. Its subtle shaping and the combination of rigidity and flexibility make a saddle a work of art and one that should last for generations.

Apart from comfort and looking good, a saddle must also accommodate the complex movement of the horse's body, for example in the trot when the horse's back extends, flexes and rotates. If these movements are restricted by a pinching tree or pain from localised pressure, the horse's gait will be directly affected.

The saddle's rigid inner structure or 'tree' protects the horse's spine and withers and distributes the rider's weight evenly. The larger the area of the saddle in contact with the back, the lower the average pressure on the horse. However, comfort for the horse also depends on many aspects, including having the correct gullet width, the angle of the panels in contact with the ribs, padding round the points of the tree, and fore and aft stability. There is general agreement that the saddle should not extend backwards beyond the 18th rib because the anatomy is not sufficiently supportive and the muscles over the loins need to move freely up and down, particularly at the canter and gallop.

Saddle trees are made of wood, polymer, carbon fibre or metal, each of which has different properties of weight, strength and flexibility. A universally-used construction of a relatively flexible tree, made of glue-bonded veneers of beech or birch was patented by Len Holmes in Walsall in the 1960s. The veneers are moulded by heat and pressure in a vacuum press to make a laminate which conforms to the required shape and size of tree. The desired cantle shape is attached by glueing and riveting.

Webbing straps, padding and finally leather covers are attached to the tree. The padding of the panels between the tree and the horse's back is a matter of choice: the usual materials are foam or flocking either synthetic or natural (sheep's wool, preferably merino), and air which can be introduced through a valve into sealed compartments. All types should be adjusted at regular intervals to check the fit with the horse's back. A good saddler can tell quite a lot about the owner's riding style from the state of the flocking in a saddle. Just as some people walk with a soft footfall or a heavy tread, some sit lightly in the saddle or ride like a sack of potatoes – it depends

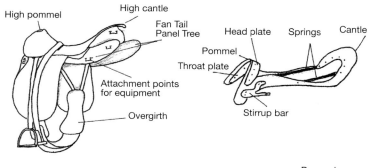

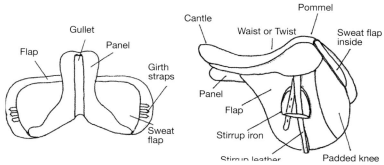

Top left, Universal Pattern (UP) British Army saddle (after C.G. Trew), general purpose saddle tree with sloping head (top right), underside of general purpose saddle (bottom left) and general purpose saddle (bottom right).

on how they control the distribution of their weight through core muscles.

Metal reinforcing plates are riveted to the head of the tree (forearch) both above (head plate) and below (throat plate) to prevent it from spreading outwards. The head may be vertical (strongest), sloped or cut back depending on the type of use and the horse's conformation.

Springs (flat steel strips) can be inserted between the waist and the cantle of the tree to improve the comfort for horse and rider. However, they can break so are not usually included in polo or hunting saddles which need to be especially robust.

Synthetic polymer trees are formed from liquid resin injected into a mould which is machined by computer-aided design to match the conformation of the horse's back. The head and throat plates are spin riveted to the tree, a process which avoids fracturing the polymer. A foam seat is glued into position and the tree is sent to the saddler for completion. The points of the tree may have leather caps sewn on them to relieve localised pressure, particularly when they are long, for extra stability.

The stirrup bars (fixed or adjustable) are also riveted to the tree. Their position determines the leg position of the rider and so varies according to use: for example, forwards for racing and jumping and central for a straighter leg in dressage.

Each equestrian discipline interprets saddle construction differently: for example, a cavalry or Army Universal saddle has a long padded fan-tailed tree extending backwards under the saddle seat which provides a large surface area. This worked efficiently during military campaigns where much weight was carried for many hours at a time. Similarly, Western saddles cover much of the horse's back and are good for a horse and rider spending long days on the hoof. In contrast, a tiny race saddle may weigh just a pound or two but although the jockey stands up in the stirrups most of the time, his weight must still be carefully distributed to allow the horse's back muscles to work freely.

Saddle-making and fitting is a highly technical and skilled subject. Progress in design is being made with pressure-sensitive computerised pad technology which can demonstrate areas of localised pressure under the saddle while the horse is in work. This type of pressure may interrupt blood flow through surface capillaries and affect the muscle function by causing pain and at worst, cramping. Badly-fitting saddles are a likely cause of lameness, according to research carried out at the Animal Health Trust's Centre for Equine Studies.

The correct fit is paramount, whether or not the saddle has a full, half tree or even no tree.

A well-controlled trial in 2017 (Murray et al., 2017) compared the pressure distribution of different dressage saddles using Pliance technology. The results showed that the design of saddles can enhance or hinder the horse's gait (degree of knee and

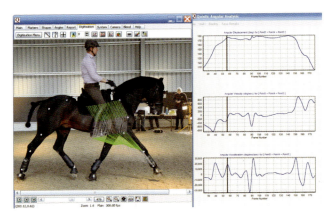

Part of a trial to test the effect of saddle and girth design on the horse's gait with pressure sensors (white dots) and video analysis.

cantle, the extent of padding of the saddle panels either side of the spine and the shape of the head of the tree. Put simply, advanced measuring technology has stimulated innovative saddle designs which produce an improvement in the horse's performance. It is, of course, more complicated in practice as a saddle is not just chosen to fit the style of the discipline, i.e. a straight flap for dressage and a forward cut one for jumping, but it is how that style fits the anatomy of the individual horse's shoulder angle and the curve of its withers that is important.

Adaptations of saddle styles for various equestrian disciplines are described in the relevant chapters.

Girths

The position of the girth straps (billets) is important for the stability of the saddle (apart from actually keeping it on) and depends on the conformation of the horse. Arabs and continental warmbloods tend to

hock flexion) and therefore performance. The crucial design elements which reduced localised pressure and improved performance were: the re-positioning of the front (point) girth strap away from the point of the saddle tree to a little further back towards the

Happy Days. Lady Violet Munnings riding side saddle on 'Magnolia' with her dogs on Exmoor 1924.

have straighter shoulders than the more sloping ones of a thoroughbred or Irish hunter. Girths further back suit the thoroughbred while a more central girth position stabilises the saddle on a horse with more upright shoulders. Wrongly positioned girths can drive a saddle against the horse's shoulders and interfere with its stride.

Many general purpose saddles have three straps for the attachment of the girths both for reasons of safety and for finer adjustment of the saddle on the horse; the first (point strap) is sewn to internal webbing behind (caudal to) the point of the tree while the rear two are both attached to a separate webbing strap within the saddle. Conventionally the girth is buckled to the first and last girth strap.

Dressage saddles usually have two girth straps which buckle below the saddle flap to reduce bulk under the rider's leg. The point strap is often attached to the point of the saddle tree but this can lead to localised pressure (*see above*), though it should be possible to pad or flock the point to prevent this. Saddles which avoid this design by attaching the girth straps caudally to the tree point have been shown by gait analysis to enhance the horse's performance (Murray et al., 2013).

Until 2013 no detailed study had been made of the pressures of the girth on the horse. Elasticated (rubber and cotton webbing) girths were assumed to be more comfortable because of the stretch to accommodate the expansion of the horse's ribcage. In fact, pressure mapping has shown the opposite to be true; the ribcage expansion in the region of the girth is minimal compared with further back behind the saddle so elastication is unnecessary. Also, elasticated girths are difficult to tighten correctly. What really matters is the accommodation of the skin and muscle between the horse's elbow and the girth. As the foreleg moves back, it pushes the soft tissue against the hard edge of the girth with every step. Maximum pressure was recorded in this area which is also a common site for girth galls or sores due to pressure and friction. A girth designed with a curve to avoid the area behind the horse's elbow showed that the horse had an increased stride length compared with other designs tested.

Side saddles

The side saddle, sometimes called 'the Queen of Saddles' has a well-documented history of development. The Goddess of the Horse, Epona, is often depicted on a horse sitting sideways to the right or offside, as shown in British archaeological finds of the 1st–4th century AD. The Queens of France traditionally also rode aside on the right, perhaps because a mounted guard alongside could draw his sword in her defence more easily. The vast majority of side saddle riders sit with their legs on the nearside of the horse, though saddles for offside riding are available.

In medieval times, women rode either astride (Eleanor of Aquitaine campaigned astride with her husband Louis VII in 1146) or, more usually in Britain and certainly for the aristocracy, sat on a pillion pad behind a man riding astride. Anne of Bohemia (1366–1394), the wife of Richard II, is said to have introduced a style of side saddle on which the lady sat on a padded seat facing sideways with her feet resting on a wooden panel or planchette, while her palfrey or horse was led. A heroic journey is told how, in 1566, Mary Queen of Scots who was six months' pregnant

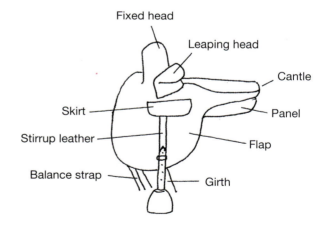

A modern side saddle. The design is credited to Jules Pellier for the leaping head and stirrup in 1830 but contested by François Baucher of Franconi's Circus and even earlier (before 1820) by an English huntsman, Thomas Fitzharding-Oldaker.

Part of the collection of Roger Philpot, international side saddle judge, instructor and fitter. These saddles are 19th and 20th century examples from many different countries.

Laura Dempsey, Master Saddler, side saddle specialist, in her workshop. Her prizewinning saddle is displayed (on the saddle horse on the right).

with James I, rode pillion behind Lord Erskine for 25 miles during the night to escape the Protestants.

The modern European side saddle developed from the 16th century. Catherine de Medici (1519-1589), a formidable rider who hunted all over France until she was sixty, is credited with hooking her right knee round the horn on the saddle and facing forwards instead of sideways. She developed a hunting saddle with two central horns, in between which the right leg could be wedged. Her contemporary, Elizabeth I who was also a keen and courageous huntress is shown in portraits riding aside and holding on to a central horn facing three-quarters forward. One of her saddle covers is displayed in Saddlers' Hall, London.

In the 19th and 20th centuries Mayhew, Champion & Wilton and Owen were the three major side saddle makers in London, all using the fixed and leaping heads design and their refurbished saddles are still being used today. Each company patented their own safety stirrup bars and Champion & Wilton introduced the balance girth in 1880 to avoid a tendency of the saddle to twist and to give it greater fore and aft stability.

Early side saddles had a tree of solid beech wood carved to the correct shape but later trees are made from birch veneers glued and vacuum pressed into a laminate conforming to the shape of a metal mould, as for the trees of astride saddles. The birch comes from Finland which provides a steady supply of wood with the right growth rate and moisture content. The wood is sealed (scrimmed) to prevent warping, webbing is attached to contain the wool flocking, and steel supports for the stirrup bar and pommel are riveted onto the tree. A cover of wool serge (or linen or leather) is stretched and stitched over the flocking and the leather built on top.

The challenge for side saddle makers is to accommodate the asymmetric weight of the rider in such a way that both the horse and rider are comfortable. The tree is modified and has a long point on the nearside and a shorter one on the offside. This long point and its surrounding flocking should closely match the curvature of the horse's side to provide grip and stability.

Flocking is slightly reduced in the seat panel on the offside to shift the rider's position to that side and counteract the weight of the legs.

A serge underside is warmer than a leather panel, grips better and is easily adjusted. Alternatively, a Wykeham Pad, patented by Mayhew in 1903, can be used under the saddle. This is a felt pad stuffed with wool and attached to the tree by two or three buckled straps. Wykeham Pads were an economical way for an owner to fit out several different horses with just one side saddle and each horse would have its individually balanced pad. This was also very practical as wet pads could take at least a day to dry out and a dry pad could be easily substituted so that no riding time was lost.

Side saddle riding is the ultimate expression of grace and modesty but is certainly not limited to hacking in the park. Point-to-pointing, hunting and show jumping are all practised by aside riders and a world side saddle show jumping record of 2.07m over a puissance wall was set by the Irish rider Susan Oakes in 2013. Side saddles for each of these uses are modified: for example, a half tree for racing and a leaping head with a rounded edge to accommodate the leg in the forward seat for jumping. Many of the modifications are shown in the outstanding collection of Roger Philpot: a wooden 'stirrup' on

A record side saddle-only meet hosted by the Quorn with more than fifty riders from several European countries, the youngest being Lucy Burton aged 12. Shirley Oultram (centre on 'Wish') and Susan Oakes, high jump record holder are among many expert side saddle riders pictured. At least six hunts now organise regular meets every season.

an Icelandic side saddle for the extreme cold and an Indian one with ventilation holes along the side so air could circulate through the panel for the hot climate. There are several different ingenious quick-release designs of stirrup bars and stirrups and Roger has copies of the original patents. The collection is displayed in the house originally built by a Mr Puckle who was responsible for training remounts in World War I. He designed the Kineton noseband, a drop noseband attached to the bit which focuses pressure on the nose whenever the reins are tightened, which he recommended for forward-going horses.

Laura Dempsey, an award-winning Master Saddler, specialising in side saddles, has her workshop near Kineton in Warwickshire. This is a historic site where the English Civil War saw its first serious engagement in the Battle of Edgehill in 1642. Two superb saddles for men riding astride, which survive from this period and perhaps this area, are referred to below.

Laura makes modern side saddles and refurbishes old ones. She qualified through the Society of Master Saddlers (SMS) City and Guilds scheme at the Saddlery Training Centre, Salisbury. After this she joined Richard Godden who worked for Champion & Wilton, London, saddle makers who continue to hold the Royal Warrant originally granted by Queen Victoria. Laura has won the Side Saddle Association Trophy five times at the annual SMS competition as well as the Best Overall Entry in 2008. She is now a judge and saddle fitting examiner for the SMS, to which she was elected president in 2015. In recognition of her work, she was made a Freeman of the Worshipful Company of Saddlers and given the Freedom of the City of London.

The side saddle trees from the 1920s and 30s are still used in today's refurbished saddles and not until recently have any new trees been made. Bespoke additions found on old saddles include a fold-away stirrup under the offside flap for the groom to ride astride to the meet and a small offside leather pocket, not for a hip flask, but for a lace hanky. These heirloom saddles spell out their age during refurbishment in the form of dead moths, mice and dust but after the leather is repaired and the panels are re-flocked they are ready for another few generations to enjoy.

An elegant lady riding side saddle wearing a silk hat, veil and immaculate habit certainly has the 'wow' factor! Today's outfits are quite demure compared with the extravagance of those worn by Elizabeth, Empress of Austria for hunting with the Pytchley when visiting England in 1874. Over a bodysuit of chamois leather, she wore a tight-fitting blue jacket with an astrakhan collar and gold buttons, just one of sixteen habits she brought with her. Her flowing skirt reached to the horse's hocks and she wore a brimmed hat with a large ostrich feather.

Side saddle riding peaked in the early 1900s then declined by the 1950s due to the expense as well as the emancipation of women during two world wars who found that riding astride was generally more practical.

It was naughty, sporty Coco Chanel (born Gabrielle Bonheur, 1883–1971) who used her fashion designer skills and flamboyant lifestyle to campaign for women's liberation and popularised jodhpurs for women to ride astride. Until then, women's clothes had been physically restrictive and trousers were

This full-size, 'winged' saddle in red and yellow Porosus crocodile, orange Swift calfskin, red and green lambskin was commissioned from Hermès by a Japanese pop group for their album cover. It was the centrepiece of an exhibition 'Leather Forever' in London in 2012 which also featured their Talaris saddle with a synthetic tree designed for show jumping.

only for men. In her early twenties Chanel became the mistress of Étienne Balsan and moved from Paris to his chateau in Compiègne. There, among the hunting social set, she learnt to ride, initially borrowing Étienne's shirt and trousers and causing a sensation. She carried off this attire with such élan that she soon set the trend for ladies to ride astride in matching jodhpurs and jackets which she designed herself.

Controversy rumbled on though. In *Mount and Man* a book published in 1927 by Lieut–Colonel M.F. McTaggart, DSO and with a foreword by Field-Marshall Viscount Allenby, GCB., there is a chapter 'Should Ladies Ride Astride?' After reviewing the aesthetic and female anatomical aspects it did however conclude that there was no good reason why ladies should not ride astride.

There was also concern about children riding side saddle, as a matching pair of offside and nearside pony side saddles in the Royal Mews shows. It was thought that a child would develop a crooked spine by only riding on a nearside saddle so the offside saddle was made to be used by the child on alternate days.

Side saddle riding is now enjoying a revival, both in the UK and the USA (where BBC dramas such as *Downton Abbey* have influenced the trend) and at least six hunts across the country hold 'side saddle only' days each season.

The Side Saddle Association was formed in 1974 and has over 1,100 members worldwide. The Association covers the UK and sets the standards for competitive events, holds teaching clinics and advises on all aspects of side saddle riding (www.sidesaddleassociation.co.uk).

Every type of saddle is a work of art – as well as an art form – as demonstrated by Hermès (*see* illustration opposite), whose expertise in leather work began in 1837. The company started as a harness maker in Paris and now makes highly technical saddles for Olympic equestrian teams. As well as their diversification into luxury goods, they continue to make top quality saddlery.

Antique saddles

Wherever the horse was important to society in times past, the saddle (and to a lesser extent, the bridle) was an indicator of its owner's wealth and prestige.

The saddlery of royalty is particularly splendid; Henry V's saddle is in Westminster Abbey (Library) where his funeral was held in 1422. The canvas seat cushion is stuffed with hay and covered with a cloth of purple velvet decorated with powdered gilt fleur de lys. Under the seat it has sidebars of hardwood joined by a wooden arch front and back to form a high pommel and cantle. A recently restored green velvet saddle cloth with gold lace and fringe was reputedly used by Elizabeth I on a visit to Bristol in 1574 and is displayed in Saddlers' Hall, London, while a 1651 inventory of Charles I lists saddle cloths of richly embroidered velvets decorated with pearls.

Queen Victoria's saddle made by Gordon & Co., London. It has a quilted pigskin seat and panel and was first used by her for a review in 1856. It has fixed heads more similar to an 18th century design and no balance girth. The stirrup, not shown, was a 'slipper stirrup' with a fixed velvet-lined, covered slipper for the Queen's left foot. With such an insecure seat, the Queen would not have had full control over the pony and was led by her servant, John Brown, when riding at Balmoral, as depicted in the 1997 film *Mrs Brown*.

Personalised decorations on the saddle leather can be seen in 16th century Tartar saddles (Ashmolean Museum, Oxford) and Victorian quilting was common. Modern Western saddles are still very decorative, often with bespoke tooling.

By the beginning of the 20th century, quilting fell out of fashion and was thought of as too ornate. It was retained for saddles exported to the colonies though, for the functional reason that the unevenness of the surface allowed a better airflow and comfort for riders in hot climates. Today's saddles are the least decorated in their history compared with the exuberant expression of past saddlers but modern tack is much more practical and easy to maintain. In the view of Master Saddler Nick Creaton, ornate saddles were not so practical for hunting and the trend towards plainer saddlery may have started among the hunting fraternity in Yorkshire.

Two fine examples of 17th century saddles are safely conserved in museums (below and previous page). When new they must also have been kept in a tack room and it is interesting to speculate what it might have looked like in the reigns of Charles I and II. The saddles would have been cared for by a hierarchy of stable yard staff in tack rooms that have probably altered little. Bridles and saddles would have been more ornate and bits potentially more severe with long shanks. Buckets were often made of leather, as at Calke Abbey, and, on the tack room shelves, there would have been fearsome tools for dentistry, dosing horses and docking tails.

An exquisite Welsh saddle (Walsall Leather Museum) has a quilted doeskin seat with botanical motifs such as tulips embroidered throughout in silvered thread. It is possible to speculate that the tulip motif may itself suggest the approximate age of the saddle. 'Tulip mania' began in Holland when tulips indigenous to Turkey and the central Asian steppes arrived in Antwerp in the 1560s from Constantinople (Istanbul). At the peak of the craze, canal-side city houses were being exchanged for single bulbs and tulips were the most fashionable flowers in European gardens. Tulips are recorded as being grown in England as early as 1611 by John Tradescant before their popularity spread more widely. Sir Thomas Hanmer (1612–1678) was a renowned tulip grower in his Welsh estate in Flintshire and perhaps the saddle came from there. The stable yard at the Hall has long since been converted.

A wave of Huguenots to England in the 1680s encouraged a British mania for tulips which lasted until about 1710. If the tulip motif was the contemporary fashion, then the saddle could be consistent with having been made in the 17th century. During the interregnum of 1649 to 1660, Cromwell forbade woodcarvers from decorating rood screens in churches and some turned their skills to carving chandeliers and perhaps to creating patterns on leather. The mystery of the origin of this saddle adds romance to its beauty.

Saddle with doe-skin seat thought to be 17th century. The style is broadly similar to those used by William Cavendish, Duke of Newcastle. This saddle was found in a Welsh barn but nothing is known of its provenance.

The saddle may also have had an upholstered seat attached for a lady to ride aside behind her knight. These pillion seats were often of quilted doeskin and stuffed with deer hair over a rudimentary wooden frame. Some have an iron hoop handle attached and a foot-rest (planchette) suspended from them.

Another rare and beautifully decorated 17th century saddle, with a lightweight wooden tree, a

The late Mark Roberts in his Aladdin's cave of a work shop. The work of saddle repairers is indispensable to the equestrian community and keeps our tack safe and serviceable.

leather seat covered with layers of quilting and a cover of turquoise blue silk velvet edged with braid, is in the collection of Banbury Museum. It had been hidden behind a wall on the Stanton Harcourt estate in Oxfordshire and only revealed two hundred years later during building work. This was a Royalist stronghold and perhaps, in the heat of the Civil War, its owner hid the saddle there, intending to retrieve it later. Brackets on each side of the front of the saddle held a matching pair of finely embroidered flintlock pistol cases.

Development of the saddle

The origins of saddles with trees and the first use of stirrups are vague. Their development was influenced by the ebb and flow of colonisation and constant interchange along trade routes.

The very earliest beginnings of saddles that we know of are Scythian woven saddle cloths and a felt and leather pad saddle about the 7th century BC. The designs and workmanship are sophisticated and beautiful. The Pazyryk rug, with decorative borders of horses wearing saddle cloths, was found in a burial site in Altai, Siberia and is now in St Petersburg. Any precious stones and gold in this grave had been looted by robbers who, by damaging the access to the tomb, allowed water to enter. The water froze and perfectly preserved the rug and remaining artefacts for several centuries.

Before 1st century BC

Pad saddles were used at least until 1 AD and consisted of a heavy saddle cloth secured with a girth. Some of the Assyrian cavalry in the 1st century BC are depicted in reliefs riding on heavy cloths secured with straps round the chest and tail to stop it slipping, as the forerunners of the breast strap and crupper.

Rolls of padding were stitched to the front and back of the pad saddle as a rudimentary pommel and cantle. Pad saddles are depicted on a column relief in Constantinople and a bronze figurine (338 BC) of Alexander the Great on Bucephalus. In some central Asian graves from the 5th and 4th centuries BC saddles with a padded wooden frame have been found.

1st–4th century AD

A Roman saddle had a horn at each corner and no stirrups. Historical re-enactment societies have shown it provided a fairly stable seat for a rider, even without stirrups. It could probably also be adapted to be a pack saddle. Reconstructions from incomplete archaeological remains suggest that the horns were made of bronze and possibly wood with a leather seat. Written accounts in the 12th century describe the cavalrymen of Theodosius I, who ruled from 379 AD to 395 AD, as having a true saddle with a tree, pommel and cantle, though no stirrups. The Venerable Bede (673-735 AD) wrote that the English started to saddle their horses in 631.

6th century

Saddles with sycamore and pine frames were found in burial chambers in Germany and Italy. These had high pommels and cantles, sometimes bound in metal and richly decorated with precious stones, enamelling and gold. The frame had no tree points either side of the withers for stability and required a second rear girth to secure it.

There are several possible origins for the stirrup. Stirrup-like loops of leather on a Scythian saddle are depicted on a Greek vase of about 400 BC and metal stirrups were used in China in 3 AD. Nomadic steppe horsemen, the Avars, are thought to have introduced the stirrup to Europe in about the 6th century as well as influencing the style of saddles.

11th–12th century

Saddles with wooden side panels connected with arches at the front and back to make a high pommel and cantle are shown in the Bayeux Tapestry. The riders have long stirrup leathers and stirrups – this was the first evidence of their use in England.

16th century

A saddle excavated in Southwark (now in the Museum of London) had a beech wood tree, iron rivets, bands and plates and the seat stuffed with horsehair.

17th century

Saddles had a wooden tree, knee rolls and a semi-circular cantle with expensive upholstery as used by William Cavendish, the Duke of Newcastle. (*see* Dressage, page 94).

19th century

The English hunting saddle. Much flatter and the seat was of leather, sometimes quilted. The wooden tree was padded with wool flocking.

20th and 21st century

The Army Universal Pattern saddle of 1912 had a long, fan-tailed tree to spread the weight, and an English hunting seat. The basic design is still used by the military.

Modular saddles can have adjustable trees made of laminated birch or beech, metal, carbon fibre or synthetic plastics. The choice of padding is traditional wool flocking, artificial fibres, foam or air. The style, particularly the cut of the flap, varies depending on the activity.

Specialist saddles

These were designed for a particular purpose, such as for jousting in the Middle Ages which had to protect and support the rider against the impact of a lance. They were often finished in finely chased metal and had high pommels and semi-circular high cantles.

Pillion pads for ladies to ride aside, either behind a rider or alone, had a padded frame.

The pack saddle of ancient origin, and vital for transport before canals, roads and railways, was used during World War II and is still important in many parts of the world.

CHAPTER FIVE

Bridles

Bridles are devices for a rider or driver to control the horse through pressure on the outside of the mouth (lips), nose, jaw and poll as well as inside on the palate, tongue and bars (*see* diagram on page 30).

Usually the focus of the pressure is a bit or a noseband and this is needed in some degree to control the horse. However, a bridle is a connecting set of levers and the referral of pressure, for example to the poll area of the head, caused by the action of the bit has only been assumed from the horse's behaviour, such as the lowering of the head. But now there is a way to measure pressure quantitatively. The Pliance system provides precise information on the localisation and the amount of pressure. It comprises a thin pad containing pressure sensors which is positioned under parts of the bridle to record data while the horse is being ridden.

Rupert and Vanessa Fairfax (Fairfax Saddles, Walsall) are saddlery manufacturers specialising in innovative designs using traditional leather as well as synthetic materials. Their aim is to translate new data from the Pliance system into their products for different disciplines; both are also experienced riders, Rupert in polo and Vanessa in international show jumping.

Vanessa is qualified in industrial design and technology and she is a Society of Master Saddlers saddle fitter and course assessor. Her approach has been to measure the pressure localisation of parts of the bridle such as the headpiece, browband and noseband and then to design a way round them. She achieves this by padding key sections of the bridle with Prolite foam inserts to lift the strap away from sensitive areas of the horse's head. In addition, the symmetry (placement of buckles), stability (fixed rather than loose browbands) and balance of the bridle are also important considerations.

Computer analysis of the data showed that bridle design can influence the performance of the horse in an unexpectedly fundamental way. Fairfax combined pressure mapping tools with gait analysis to study horses while they were being ridden. Gait analysis is photographic high-speed motion capture (about 400 frames/sec) – reminiscent of Eadweard Muybridge's 19th century experiments – using inertial motion sensors placed on the horse's legs and body.

The results, published in veterinary journals, showed a common trend: that the relief of pressure on the horse's head increased the power of its stride. Analysis showed significantly greater flexion of the knee and hock as well as an extension of the hind legs further under the body. The improvement shown in the horse's performance following bridle adjustment is immediate and several owners also reported better general behaviour as a bonus.

Master Saddler and Master Bridle Maker Issi Russell specialises in the design and making of fine leather bridles. Her clients include the 2012 Olympic Gold medal winners (team and individual dressage) as well as professional and leisure riders.

Issi has craftsmanship in her genes, her grandparents were coachbuilders and glovers and, inspired by a visit to Saddlers' Hall in London, she went on to complete the Cordwainers' Diploma and

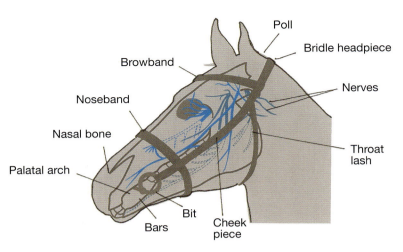

A diagram of the distribution of nerves in the horse's head and how the bridle may cause pressure.

the Higher National Diploma in Saddle Technology at Capel Manor College, London. The college has been renamed The Princess Royal College of Animal Management and Saddlery and offers the only full-time saddlery course currently available.

This was followed by two years' apprenticeship to Laurence Pearman during which she won the Les Coker Millennium Trophy for the best entry by an Apprentice and the Best Bridle award. In 2008 she became a Master Saddler and other awards followed including the Highly Commended Certificate for the Prince Philip Medal. Issi has been a national assessor and judge, an executive committee member of The Society of Master Saddlers and self-employed since 2013.

Working in the traditional way she brings her own innovation to the design of bridles. The demands of international dressage riders are considerable. During the dressage test, often held in indoor arenas in a highly charged atmosphere, any discomfort from the bridle would cause tension in the horse's jaw and neck and lose marks from the judges. Her work is by commission only, for which she chooses Sedgwick's (Walsall) top quality leather. Issi's designs are sympathetic to the sensitivity of nerve plexuses particularly round the ears so that the horse is comfortable and fully able to concentrate.

The headpiece of one of her double bridles is shaped to curve well behind the ears to give them more ease and its extra breadth distributes pressure from the bit and reins. The noseband is usually held in place by a narrow strap which lies under the bridle headpiece behind the horse's ears. Instead, Issi has inserted the strap into the bridle headpiece padding. Secondly, she has brought the strap carrying the snaffle bit (bradoon), known as the bradoon sliphead, outside the bridle headpiece so that it lies on top. This arrangement means that the wide headpiece, now in direct contact with the skin, distributes pressure more evenly over the poll.

The insides of the loops of both ends of the oversize browband (and noseband) are shaved (skived or feather-edged) to lie as flat against the head as possible and strength is maintained by the invisible insertion of light webbing. The noseband (cavesson) is padded across the nose and under the jaw and can be of natural or patent leather. The padding is made from soft but hardwearing foam which has no 'memory'. Design doesn't stand still though and Issi is constantly working on improvements.

Patent leather is popular for bridle nosebands as well as for dressage riders' boots. The hard shine of patent was originally a Japanese process (japanning) in which the leather was varnished with plant resins and shellac. All patent leather now is made by bonding a plastic coating onto leather. It has the advantage of not cracking with age, as japanned leather did, but it scratches easily.

Variations of bridles include bitless bridles and hackamores which rely on exerting pressure on the head and nose. Bridles with blinkers (winkers) around the eyes for driving or racing restrict vision to a forward view. Agricultural and cavalry bridles are made of heavier leather (4.5mm) than riding bridles (3.5mm). Western riding bridles are more ornate, have no noseband and often a loop over only one ear to hold them on instead of a normal browband which is attached to a headpiece behind both ears. Military and endurance bridles are a combination of headcollar and bridle in which the bit and reins can be easily detached to allow the horse to graze or drink.

Master Saddler Issi Russell of IR Bridles in her light and modern workshop. The browbands (on the centre wall) can be customised to suit the owner – perhaps braided in colours to match an owner's racing silks or decorated for showing to catch the judge's eye with diamanté, glass beads or pearls.

Reins

Reins are usually made from leather strips either side of the spine line of the hide and a standard length is sixty inches. Reins with a variety of grips are available depending on their use: plain, plaited, leather laced or rubber sleeved. All are buckled at their ends to make a continuous loop except for Western reins which are usually kept separate (split).

A horse having its computerised pad (blue) under the noseband and bridle sensors (yellow dots) adjusted during Fairfax pressure mapping trials.

Martingales

There have been many designs of martingales which are used to prevent the horse from raising its head beyond the line of control i.e. beyond the point when the leverage action from the rider's hands on the bit is effective.

Martingales may be worn for steeplechasing to keep the horse's head in a position to fully view the jumps. They can also prevent the back of the horse's head hitting a forward-leaning rider on the nose or they can be used for schooling an unmounted horse on the lunge.

The two most common types are the running martingale, a strap attached to the girth which divides to connect to each rein with a free-running ring, and a standing martingale which is a strap running from the girth to the underside of the noseband. Both types have a neck strap to keep them in position. The rider has some control over the action of a running martingale by using the reins but a standing one is independent of the rider's hands and is usually used for polo and polocrosse.

Nosebands

Nosebands are of many designs and their action is crucially dependent on their adjustment. Some nosebands exert pressure on the lower end of the nasal bone and around the mouth and can be severe.

Above: A bridle strap held in a saddler's clam (vice). The leather strap is rolled round an internal cord for strength and its edges are held together by an awl during handstitching with linen thread which has been drawn through a block of beeswax.

Above: The finished double bridle during a fitting on Donita (owned by Helen Wicks) – the curb chain to be added later.

Left: An Issi Russell double bridle in the making on the workbench. The narrower bradoon sliphead strap is shown running over the top of the padded headpiece.

For Bits – *see* Lorinery and Bits, page 41

A cavesson, a simple strap noseband held well above the horse's muzzle and away from the bit, has the least effect in terms of pressure, provided it is not fitted too tightly. Drop nosebands, where the strap round the nose is positioned in front of the bit, not only keep the mouth closed but affect the action of the bit itself. Various arrangements of straps in front of and behind the bit, such as Flash or Grakle nosebands (each called after a horse) are a matter of individual choice and their exact modes of action are becoming clearer due to the computerised sensor systems described above.

CHAPTER SIX

Harness

Terry Davis is a well-known master harness and collar maker and has a long-established business near Ludlow, Shropshire. His top of the range heavy horse harness is used throughout the country and there is continuing demand from the UK and Europe. However, a recent report on traditional skills by the Radcliffe Trust found only three horse collar makers in the UK and concluded that this is a critically endangered skill.

Terry was brought up in Northern Ireland but left for London when the 'Troubles' worsened in the 1960s. He took a trainee position with George Parker & Sons, a prominent saddle and harness-making company in St Martin's Lane and attended courses at the Cordwainers' College in Hackney before setting up his own business in Twickenham. After a holiday in their faithful VW camper van in Shropshire and a chance visit to Acton Scott Historical Working Farm, Terry and his family moved to the area in 1983 and harness and collar-making became the focus of his business.

Terry's workshop is highly traditional where nothing is computerised and everything is done by hand with the help of two sewing machines, one a Singer No.6 which is a 120 years old. There is a bewildering array of equipment: leather clams which are placed between the knees to hold work while hand-stitching and moulds for making bridle blinkers where the wet leather is fitted over the mould and put in a press. Also on the long bench are several pricking tools which mark out on the leather the required number of stitches per inch, as well as an assortment of cutting, edging and sealing tools.

Terry Davis using a stuffing iron to align the rye straw in a new collar. The woollen collarcheck is pulled up round the straw before being encased in leather on the outside. Rye, picked green before the seedheads develop, has tough but pliable stems which do not disintegrate easily. A wet leather sleeve is moulded round the body of straw, spot stitched and beaten into shape before being handstitched into place.

Fifield Admiral (Monty), winner in 2017 of the Harness Class at the National Shire Horse Show for the fifth year in succession. He is owned by Mike and Faye Bottomer and shown by Charlie Lloyd (ex-Head Coachman at the Royal Mews). Monty's harness was made by Terry Davis.

The leather he uses for harness-making is the finest from J & F.J. Baker in Devon (*see* page 2) and is about 4mm thick to provide the strength required. On floor-to-ceiling shelves are dozens of empty old marmalade jars used to store a variety of buckles, rings, terrets, etc. – like a saddler's sweet shop.

One of the top world suppliers of these metallic components of harness (lorinery), is Abbey England, holder of the Royal Warrant and based in Knutsford, Cheshire. Abbey also owns a foundry in Walsall where high quality brass or alloy buckles, horse brasses and other fixings are cast in traditional greensand.

While we talk, Terry is easing a twisted hemp thread to and fro over a block of beeswax to waterproof it and make it easier for sewing. Polycotton and polyester threads, which are cheaper and more resistant to rot, are now replacing natural linen and hemp yarns but there are pros and cons for each. The harshness of the poly threads can cut into soft leather while natural fibres are more companionable with leather and are less likely to stretch. There is a satisfaction to using harness made entirely from natural materials; leather, linen thread, Norfolk rye straw for the collar stuffing and crushed linseeds to fill the crupper roll to keep it oily soft and comfortable under the horse's tail. Terry reminds me that whole linseeds are never used in case they begin to sprout.

An English collar takes about thirty hours to make but should last at least a lifetime if properly cared for. Ideally, each collar should be made for an individual horse and not shared by others and it will take several weeks of work before it beds in and moulds itself comfortably to the contours of the horse's neck.

I leave the workshop reluctantly to the steady rhythm of the sewing machine foot treadle stitching a new harness for another lucky owner.

Terry's name is also known in Central America, Egypt and Africa where his research into harnessing methods was funded by a Winston Churchill fellowship award. His voluntary work abroad, among communities who depend on the family donkey, pony or mule for their livelihood, is focused on producing effective and inexpensive harness for working equids. Harness-related injuries are common but preventable when the design and function of the harness components are properly understood. Girths of wire and abrasive nylon, together with misunderstanding about how to balance loaded carts, lead to an immense amount of unnecessary morbidity among draught animals.

Workshops to address these practices are held by many charities such as SPANA, the Brooke Hospital, the Gambia Horse & Donkey Trust and World Horse Welfare. These teach the principles of draught (the forces which govern the most efficient way to pull a load, see below) and how to design simple harness which can be improvised from local materials. Leather is often not available but sugar

Stitching blinkers. The blinkers shown here are made by 'blocking' the dampened leather in metal moulds; the recess created provides more room around the horse's eye.

bags, or coir matting padded with cotton can be good alternatives. In this way much can be done to alleviate hardship, not just for the animal but for the family members who depend on it.

At its 90th anniversary conference held by World Horse Welfare in 2017, it was acknowledged that there are one hundred million working equids which supply 50% of the traction needed in the world. An even more sobering statistic is that 95% of qualified vets only see 2% of these equids.

Collars

The wheel, the horse collar and the plough are among the most pivotal inventions which have shaped agricultural life in Britain. Full collars have been known in Britain since the 12th century. Before then, yokes or the less efficient breast collars were used for agriculture, transport and war chariots. Chariot harness with breast collars probably first appeared in the Sintasha culture about 2000 BC in the Ural steppes.

Although harness has not changed in its basic design since the 12th century, various types of collars

A fine workshop with cart saddle (left), two sewing machines in the background and a spacious workbench. Terry is cutting and measuring straps using a round cutting knife, a symbol of the Society of Master Saddlers.

A donkey in Luxor, Egypt at ACE an animal welfare charity based there. It is wearing one of Terry Davis' simply designed and locally made harness pulling a correctly balanced two-wheeled cart.

have developed in different countries. For example, at a 2017 horse loggers' meeting American, English and Scandinavian collars were all in use.

Until recently it was not possible to compare the efficiencies of these different designs and Terry Davis was the first to initiate a formal comparison. A computerised pad (Techscan CONFORMat) was placed between the horse's neck and the collar and its continuous pressure mapping readouts were monitored during a standardised trial. Analysis showed that the English collar distributed the pressure on the horse's neck more effectively while pulling a standard load than a Scandinavian, American, Czech, French or a zinc collar. This preliminary study was circulated by Wolverhampton University and further work on this important topic remains to be done.

The most common collars used in the UK for carriage and coach work are English collars and breast collars. For agricultural work, an English or American collar is used and for logging, English, American or Scandinavian collars are a personal choice. Zinc collars are for temporary medicinal use and allow a horse with sore skin to continue in light work. The collar is cast from zinc metal and the zinc ions are thought to assist in healing.

The first collar trial was designed by Terry Davis in collaboration with Wolverhampton University in 2008. A computerised pressure pad was placed between the horse and six different collar designs. The English collar shown here proved to be the best with the lowest average pressure on the horse's neck when pulling a fixed-weight load.

The standard English collar is ideal for heavy draught work. It is lined on the inside with woollen collarcheck fabric which absorbs the horse's sweat.

Scandinavian collars are light (wooden hames), leather lined, adjustable at the top and fasten underneath the horse's neck.

The Principles of Draught

A horse in draught draws a load forward by pushing into its collar. A neck collar has wood or metal hames fitted into the forewale of the collar which provide the attachments, usually hooks, for the traces which are connected to the load. The power comes from the horse's hindquarters, particularly its hocks.

The 'line of draught' should be as straight a line as it is possible to arrange with the traces between the point of attachment on the harness and the point of attachment to the load.

The greater the area of contact between the horse's neck and the collar, the lower the average pressure and the easier the load will be for the horse. The rigidity of the collar and hames helps the horse to push into it efficiently. Size, padding and fitting are crucial to avoid interfering with the horse's breathing, shoulder and withers. There should be just room to slip an adult hand downwards between the collar and the horse's throat.

A breast collar is a padded strap passed around the horse's chest and attached with a buckle on either side to the traces. An adjustable strap over the neck keeps it above the points of the shoulders and below the throat. If the traces were attached to fixed points on the vehicle behind, the shoulders would be chafed by friction from the collar as they moved backwards and forwards. Therefore breast collar traces should be

Breast collars extend back into traces connected to each end of a swingle tree behind the horse (see photo left). The swingle tree is then attached to the load through a central point which pivots to allow the horse's shoulders to move freely without chafing. This harness is made of biothane and shaped to give the shoulders more freedom of movement.

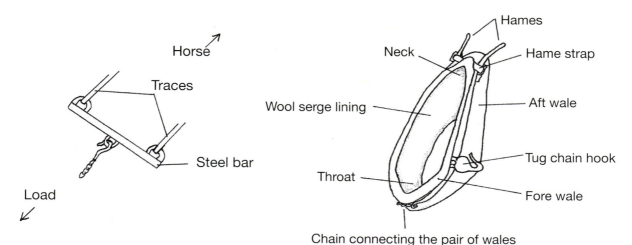

Diagram of an agricultural swingle tree. A swingle tree connects the horse to a load and keeps the traces from chafing the horse's flanks. A swingle tree for a carriage would be similar except that it would not have the attachment for a load shown but a central pivot onto the carriage.

Diagram showing the construction of a full English collar. The metal or wooden hames fit into the forewhale and are secured at the top and bottom of the collar with straps or chains. The hames provide rigidity and points of attachment for the tug chains or traces.

A donkey in Egypt comfortably harnessed as part of a harness workshop demonstration. The traces (white ropes) follow a direct line of draft from collar to vehicle. The shafts are supported by an adjustable chain running through the channel on the cart pad, a rigid frame which is padded underneath to protect the spine.

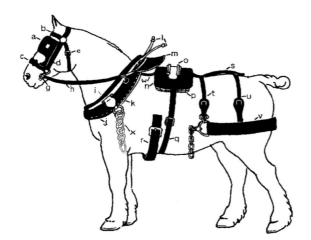
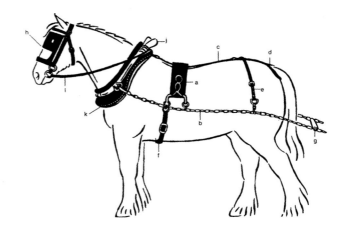

Bridle	**Collar**	**Cart Sadle**	**Breechings**
a blinker	i forewale	n saddle housing	s crupper strap
b browband	j aftwale	o bridge	t lion strap
c noseband	k padding	p pad	u hip strap
d cheek strap	l hames	q girth	v breech band
e throat lash	m housen	r belly band	
g bit	w meter strap		
h hame rein	x tug chain		

Agricultural draught harness for shafted vehicles.

a back band   g spreader bar
b trace chain   h bridle
c crupper strap   i hame rein
d crupper   j hame
e hip strap   k collar
f belly band

Trace harness (both diagrams by Terry Keegan).

connected to the ends of a swingle tree which in turn is attached through its central point to the vehicle so that it can pivot horizontally. The swingle tree swings to accommodate the horse's shoulder movement.

A breast collar can be adjusted to fit different horses. Neck collars must be carefully fitted to each horse and are generally preferred to breast collars for heavy loads as the weight is spread over the whole of the neck in front of the shoulder.

When reversing a load, the horse pushes back into the breeching strap which runs round its hindquarters. This full breeching is attached to the cart shafts by short chains. False breeching (only used on light driving vehicles) is simply a strap attached behind the horse to both shafts enabling the horse to sit back into it when reversing or going downhill.

Canadian team harness breeching is a variation on full breeching which can be used for a pair of horses working abreast with the pole of the vehicle betweeen them. Quarter straps come forward from the breeching strap on each side under the horse's belly and join into a single martingale strap running up between its front legs and attaching to either end of a neck yoke at the end of the pole.

Correct balancing of the weight of a load in a single axle, two-wheeled cart, is crucial for the draught animal. The load should never behind the axle but just forward of it. If the load is too far forward though, most of the weight will be on the horse's back.

If the load slides backwards behind the axle, the shafts will rise up and may pull up the collar to either choke the horse and even lift it off the ground. To prevent the shafts rising in this situation, a wide strap (belly band) is slung a few inches below the horse's belly and attached to the shaft ends.

The five main types of harness are:

- **Shaft harness** for vehicles with shafts
- **Pole harness** for a pair of horses abreast pulling a wagon, corn drill, mower, etc.
- **Trace and plough harness** with chain traces for horses in tandem
- **Plough harness** with leather or chain traces for single or multiple horses
- **Scandinavian harness** for shafts used for forestry work

CHAPTER SEVEN

Lorinery – Bits, Stirrups and Trees

Lorinery includes all the metal components used in saddlery and harness such as bits, stirrups, saddle trees, buckles and terrets, which are rings through which the reins or straps pass. Bronze and Iron Age pieces, some inlaid with coloured enamel, such as strap junctions for fastening chariot harness and the ornate cheek pieces on Roman bits, are very beautiful. The largest Bronze Age hoard in Britain of 93 pieces of harness lorinery was found in North Wales and is now in Cardiff Museum. This collection looks as though it was made only yesterday. The function of each component is still recognisable and, being made of bronze, has a lustrous finish unlike the rust of later Iron Age bits. It is rare for any ancient harness other than the metallic parts to survive and much of our knowledge comes from images drawn on pottery and from bas-reliefs in stone.

A ridden horse may be persuaded to change direction by neck reining as well as by the redistribution of the rider's weight, and to respond to a voice command to stop. Alternatively, bitless bridles which exert pressure on the head are used in several cultures. However, in the equestrian history of the western world, ultimate control has traditionally been through a metal bit in the mouth. This is arguably the single most important piece in all saddlery and harness.

Patterns of wear on horses' teeth which have been excavated in Botai, Kazakhstan, have been cited as evidence that horses were wearing bits before 3,700 BC but they are difficult to interpret. Leather, hemp rope, bone and eventually metal have all been used as bits. Metallic bits, either for riding or driving, leave the most distinct characteristic wear on the enamel of some premolar teeth. That the bits would have been used for riding, rather than driving, is consistent with other archaeological evidence from this settlement which suggests that these people rode horses to hunt horses (primarily for food). Also, the management of domesticated herds of horses, cattle and sheep on the Eurasian grasslands is likely to have required mounted herders from even earlier times, possibly around 5000 BC.

Among the earliest equestrian depictions of a man riding a horse holding the reins is a small Egyptian figure now in the Metropolitan Museum of Art, New York, which dates from 2000–1800 BC. Seal impressions of men riding equids (horses, onagers and asses) found in Central Asian graves are of a similar date, 2050–2040 BC.

Bronze Age jointed snaffle bits, looking very much like ours today, were cast using the greensand process which is still in use. Straight bar bits made of iron with a bronze casing and cheek rings survive from the Iron Age and tests on the metals show that trade was thriving between Britain, Ireland, the Continent and Scandinavia long before the Roman conquest of Britain in 43 AD. Greek snaffle bits with roller bars from 550 BC and twisted, jointed Scythian bits around 400 BC are examples of sophisticated Iron Age design.

The study of lorinery can tell us more about the relationship between man and horse over the last three hundred years than any other component of tack.

Part of the Bronze Age (1150-1000 BC) hoard of harness fittings found in North Wales at Parc y Meirch (Park of the Horses).

The size and severity of bits and spurs in the 17th and early 18th century are thought of as extreme by today's standards. The horse's behaviour has remained much the same so why did the bit become so exaggerated? There may be several reasons.

The horse was important as a show piece of the owner's wealth and status. The metalwork of the bridle was an ideal place to display the owner's prestige with decorations in silver, gold and enamel. The longer the shank of the bit, the greater the surface for ornamentation – and the more severe was its action on the mouth. While it may have made for a quicker response from the horse, this also suited the widely admired image of man exhibiting his superiority over the innate wildness of his steed. A draconian bit and brutal spurs that drew blood from the horse's flanks all added to the portrayal of man taming the beast.

Later, it was seen as even more impressive if the rider could tame the same beast through the power of his personality rather than his weapons of bit and spur. Perhaps with better training of the horses too, the severity of equipment could be reduced: no longer the huge ports on the bit which pressed on the roof of the horse's mouth and the long spiked rowels on the spur.

The design of the elaborate shanks, ports and curb chains of the bit, particularly in the English medieval period, may also be explained as a consequence of the need for armour. In the 14th century, to defend themselves from the enemy's deadly arrows in battle, horse and rider were equipped with armour. The restriction of the rider's movements in armour meant that he could only use relatively small movements of his arms and legs. He therefore needed the severe bits to control the horse and spiked spurs to reach the horse's flanks and urge it forwards. This theory can be appreciated by looking at the splendid collections of armoured knights (and horses) in the Fitzwilliam Museum in Cambridge and the Wallace Collection, London.

The use of gunpowder in the mid 15th century in Europe allowed a lighter chain mail to be used which gave the rider more manoeuvrability in controlling his horse and therefore less reliance on severe bits and extended spurs. Without heavy armour to carry, horses for the cavalry could be bred with a lighter and faster conformation. The breeding of quality horses led to the use of more suitable, lighter and less elaborate saddlery with saddle trees made of wood which were lightly reinforced with metal.

In the reign of Charles II, racing and hunting were popular and the style of riding with slightly shorter stirrups began to differ from the previous straight-legged, well-supported position of the dressage and cavalry riders. This new flexibility for the rider led to the design of simpler bits, bridles and saddles that were more practical.

Horses were used in most of the British military campaigns up to the Second World War, and their saddlery gradually became plainer, though the ceremonial Household Cavalry bits are probably the most ornate still in use today, together with the state harness bits at the Royal Mews. In the same way, the style of harness and saddlery in regular non-military use in Britain became less decorative, particularly since the motor car replaced the horse as a status symbol.

The National Collection of Lorinery is displayed in Horsham Museum, Sussex (*see* page 14). A more extensive collection, which includes antique and foreign examples, is at the Museum of the Horse at Tuxford, Nottinghamshire, recently founded by Sally Mitchell. It was awarded the distinction of honorary membership of the French Club International d'Eperonnerie, which specialises in equestrian antiques, in 2014 and is highly recommended.

Bits (from the Anglo Saxon word 'bitan' to bite)

The subject of bits and bitting is complex and still evolving. There is a wide variety of bits made for all types of horses and occasions but control and comfort are paramount, in fact discomfort in the mouth can make some horses pull even harder to escape the pain. Top riders, equine dentists and commercial companies all collaborate to improve bit designs so that the rider may communicate his intentions as sensitively as possible while being safely in control.

The bit in a horse's mouth fits into a natural gap (diastema) between the incisors and pre-molar teeth and rests over the tongue on the bars (or gums) of the lower jaw. The conformation of horses' mouths varies widely, depending on the breed and on the individual; some have ridged bars instead of flat ones and can be very sensitive to the bit. The Bemelman's Weymouth (made by Sprenger), used by Carl Hester, is an example of a bit which exerts a very light pressure on the bars due to being angled forwards.

The size of the tongue, the position of the teeth

High-port bits.

and shape of the roof of the mouth (palatal arch) all need to be carefully considered when fitting bits. A jointed snaffle bit with a central lozenge (KK Ultra, Sprenger and Turtle Top™ Neue Schule) suits some horses and is favoured by dressage riders Laura Tomlinson and Maria Eilberg. The straightness or degree of curvature of the bit (a port) and its thickness determine the comfort of the tongue and how easily the horse can swallow. A curb chain, which acts in the groove behind the chin and affects the action of the bit, needs to be carefully adjusted so that it just comes into contact with the skin when the shank of the curb bit is at an angle of 45°.

Different metals also have particular properties; stainless steel is very cold to the touch while the old agricultural bits with a high iron content were willingly accepted in the mouth. Just some examples of the range available are: Sprenger have patented an alloy 'Sensogen' containing copper, zinc and manganese which is thought to calm horses. Neue Schule make bits which have a high copper content and are nickel-free which can encourage salivation. Other alloys have a higher aluminium content which may have less taste. Monty Roberts and show jumper William Funnell both use American Myler 'sweet iron' bits made of iron and copper which appear to have an appealing taste and are particularly good for starting off young horses.

The Parc y Meirch hoard of Bronze Age harness fittings. The flat jangle plates (bronze discs) were worn round the horse's chest as decoration, many of the smaller pieces are strap dividers and terrets (rings through which the reins pass).

Lorinery
Bits and Bitting

Loriners Company
www.loriner.co.uk

www.beta-uk.org

Lorinery is all the metalwork for the horse's tack and harness

The Bitting Pressure Points / Inside the Horse's Mouth

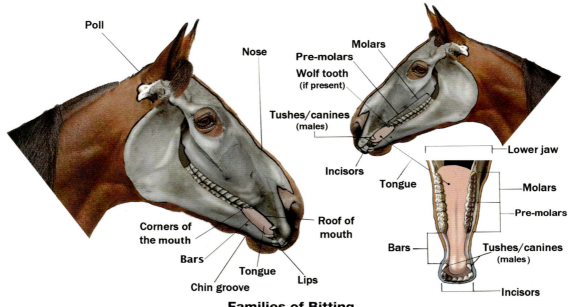

Pressure points labelled: Poll, Nose, Corners of the mouth, Bars, Chin groove, Tongue, Lips, Roof of mouth.

Inside the mouth labelled: Molars, Pre-molars, Wolf tooth (if present), Tushes/canines (males), Incisors, Tongue, Lower jaw, Bars.

Families of Bitting

BIT FAMILY	SNAFFLE	CURB (DOUBLE BRIDLE)	PELHAM	GAG (RUNNING GAG)	LEVERAGE BIT	BITLESS BRIDLE
NOTES	The most popular group of bits, they have no curb chain, poll or leverage action. Can be very mild, but if the mouthpiece is very thin or has an uneven surface, or if a double joint is twisted, this can increase the severity	This has two bits - Snaffle Bradoon and a Curb (Weymouth) which is used with a curb chain and lip strap. Horse and rider must be ready and at the correct level of training for this combination - it is not just for smart turnout	Aims to combine the actions associated with a double bridle but in one bit. The two points of attachment for the reins allow definition between the action of the upper and lower rein. Can be used with roundings and one set of reins for a more general action	These look like Snaffles but with the addition of holes in the bit's outer sections to allow special bridle cheek pieces to run through the bit, the lower rings of which the reins are attached. A useful bit for a strong horse that goes with its head too low. Often seen on horses in fast work such as polo, cross country and hunting	Often incorrectly referred to as a gag or elevator. These bits have a point of rein attachment that is below the the mouthpiece creating a leverage action. Many have long shanks but they do not have a curb chain. Can be used with two sets of reins to provide a choice of action. Bits such as Wilkie, Bevel, American "Gag" and multi-ringed bits are in this group	Not technically a bit as there is no mouthpiece, but this does not mean that this is a mild option. Particular care should be taken in the fitting and use, and plenty of time should be taken in the familiarisation of both horse and rider. Useful for a horse with a damaged or sensitive mouth
ACTION	Lips, bars & tongue	Bradoon: lips, bars & tongue Weymouth: bars, lips, chin groove, poll & tongue Curb chain: chin groove	Bit: Lips, bars, tongue & poll Curb chain: chin groove	Bars, lips, tongue and strong upwards action on corners of mouth. Some poll pressure	A small amount of lifting action as the contact is initially taken up, which is quickly replaced by downwards action on the poll, lips, bars and tongue	Nose, poll, back of the jaw (above the chin groove)
REACTION	A general upwards action to encourage the horse to raise his head and neck	The Bradoon acts as a Snaffle. The Weymouth encourages the horse to lower and flex his head and neck, encouraging a more advanced head carriage and an increased level of control	On the top rein, a Pelham will encourage the head and neck to raise. On the bottom (curb) rein the horse is encouraged to lower and flex his head and neck to encourage a more advanced head carriage. Also helps improve control	The 'running' section of the bridle's cheek pieces act in a strong upwards direction to lift the head. Can be used with two sets of reins to provide a mild to strong action. For use by experienced riders	The rein being attached below the mouthpiece results in the bit rotating when a contact is taken up. Pressure is applied to the poll and mouth with a downwards action, encouraging the horse to lower his head and neck	Pressure on the poll encourages the head to lower. Pressure on the nose encourages flexion, bringing the head inwards. NB. There are lots of different types of bitless bridles
EXAMPLES						

How to Measure a Bit — Measure here

Bit chart from The Worshipful Company of Loriners in association with the British Equestrian Trade Association (BETA)

The new pressure-sensor monitoring technology and computer analysis has an important role in lorinery design. The Neue Schule company is among those conducting research on the action of bits with the use of miniature pressure sensors placed on the bridle headpiece, cheekstraps and reins. Computer analysis of this Synchronicity™ System data can indicate where the horse may feel pressure, eg. over its poll when the reins are tightened, and helps to show where and to what degree the different designs of bits and bridles exert their force. A smartphone application can collect the data from the sensors and upload it for study later.

Stirrups

Stirrups were unknown to the Romans and are thought to have been introduced to Britain by the Normans in 1066 as they are clearly shown in the Bayeux Tapestry. In other parts of the world, a forerunner of the stirrup – a loop for the big toe – was used in India in the 2nd century BC. Written evidence of a stirrup is found in the biography of a Chinese officer in 477 AD who attributed it to the Huns and by the 7th century AD stirrups had reached Japan, China and central Asia. The acquisition of the stirrup gave the rider much more control over the horse and must have been a deciding factor in many ancient battles. Presumably, girthing techniques used on earlier pad saddles were developed alongside the stirrup so that the saddle didn't slip as the rider mounted or turned at speed.

Stainless steel stirrups have proved their reliability but there is competition from new models. Some have a steel core and a coloured synthetic exterior and polymer and carbon fibre stirrups are light and strong. There is a choice of treads: graphite for extra grip, inclined (for show jumping) or sprung (for endurance riding). Various safety designs have appeared, such as those with a caged front (not unlike the ornate antique slipper-stirrups with a covered front) which prevents the foot from slipping forwards through the stirrup. Other designs, such as Freejump, have an incomplete outer synthetic branch which would bend to release the foot if the rider fell.

A hinged safety stirrup designed to release the foot if the rider fell.

Many articulated stirrup designs were designed for use out hunting, particularly on side saddles.

Western saddles often have wooden stirrups (light and resistant to extreme temperatures), leather-covered and with an outer tough leather shield to protect the foot from thorns and to prevent brushwood from becoming entangled. The Western stirrup leather is expanded into a flap (fender) which keeps the horse's sweat off the rider's leg.

Stirrup bars, an integral part of the saddle tree, are of different designs and have safety features so that the stirrup leather will slide off backwards if pulled hard. Several were patented by the side saddle manufacturers such as Champion & Wilton, Mayhew and Owen. Many are adjustable so that the position of the rider's leg can be altered to be further forward or back.

Trees

The tree is the rigid core of a saddle which is shaped to protect the horse's spine and withers from pressure and to provide a seat which is comfortable for both rider and horse. Modern trees are made of wood, metal or synthetic materials and reinforced with metal so are within the province of the loriner.

Gibson's of Newmarket workbench showing saddles and trees. The furthest is a traditional riding saddle, then a race saddle, two wooden trees – the first one with springs, an aluminium half tree and a synthetic tree in the foreground. Metal reinforcing over the forearch and in the stirrup bars are made by loriners.

Early saddles, such as the 15th century example in Westminster Abbey library belonging to Henry V, were rigid but comprised a wooden frame rather than a tree with bars positioned either side of the horse's spine, held in place by arches front and back and strengthened by iron straps. This created a high pommel in front and cantle behind while the frame was padded on top and underneath.

A true tree is an arched structure (the head or forearch) over the withers with points projecting downwards and a rigid outline of the seat extending backwards. The tree protects the horse's spine and provides a strong surface onto which the metal stirrup bars can be screwed or riveted and from which the stirrup leathers are suspended.

Traditionally, the tree has been made of laminated wood, usually beech, glued under pressure which gives the desired combination of flexibility (to accommodate the horse's movement), rigidity (to maintain its shape) and light weight (*see* Walsall, Albion Saddlemakers page 8). The forearch (pommel) is reinforced over the top (head plate) and underneath with steel (throat plate) to preserve the shape.

'Springs' can be included in the tree; these are flat strips of steel which run from the tree waist to the cantle to lie underneath the rider's buttocks and act as shock absorbers for the comfort of both the horse and rider. As these can be vulnerable and break, polo and hunting saddles are not generally fitted with springs.

Trees of moulded polymer, aluminium and carbon fibre can be lighter in weight and titanium can replace the steel components. Half-trees for racing can reduce the total weight of a flat racing saddle to a couple of pounds.

Different forearch or head shapes vary from high to cut-back according to the intended use and the horse's conformation. High vertical forearches are the strongest and used for polo and hunting saddles. Adjustable trees were introduced by Albion in 2004 in which forearch sizes were interchangeable and seat and cantle sizes could also be changed after the saddle had been completed. Polymer trees may also be altered in the finished saddle by heat and computer-aided measurement to fit individual horses.

The manufacture of Lorinery

Abbey England, Walsall

Abbey England is a British family firm which was founded in 1982 in Altrincham, Cheshire and is now based in Knutsford and Walsall. At least three members of the family were working in different departments at the Walsall foundry on the day I visited. Abbey was awarded a Royal Warrant in 1995 as Supplier of Saddlery Workshop Materials and manufactures a vast range of high quality equestrian fittings and tools for the home and international markets. It has expanded to produce a range of equestrian rubber products, such as racing reins and overreach boots, and a bespoke fashion accessory department for briefcase locks, handbag and vintage car handles. These in-house diverse skills keep interconnecting trade channels open and production flowing against the tide of cheap imports.

The Abbey England foundry in Walsall, the last major lorinery manufacturer in Britain. This greensand casting method, unchanged since the Bronze Age, produces high quality components for saddlery and harness. Here the molten copper-zinc alloy at 800°C is being poured into the pattern moulds to make the cheek pieces for bits.

There is a world of difference between a solid brass buckle with a strong stainless steel tongue which will always look fine and last for generations and an imported cheap, die cast one on which the surface colour soon wears down to white metal.

To make a particular component, such as a snaffle bit, the lorinery process begins with the making of a 'pattern' or prototype which is a highly skilled job. One of the earliest pattern-makers recorded was Robert Plot (1680) who produced several designs of bits, spurs and curb chains, though Walsall was already a centre for lorinery by the 15th century.

The 'pattern' of a stirrup for example, is made out of metal by hand. It is pressed into moulding sand contained in a frame, or casting flask, and the lid closed. Pressure is applied, after which the frame is opened and the pattern stirrup carefully removed leaving a hollow shape matching it. Molten metal is poured into the hollow then allowed to cool. The newly cast stirrup is freed from the sand mould and finished by being 'barrelled' in a drum for 48 hours with pebbles of different abrasive grades to remove any rough edges. Several stages of hand-finishing or 'fettling' with polishing wheels, or in a ball-bearing machine, produce a shiny and durable surface – hence the saying 'To be in fine fettle'.

Until 1811, when a new type of steel lent itself to the casting process, stirrups were hand forged from red hot strips of iron hammered out on an anvil. Different casting alloys are used depending on the product. For bridle components, 70% copper and 30% zinc would be suitable; for saddle stirrup bars Aluminium Bronze AB1 made of copper, tin and aluminium, heated to 1,200°C would give the extra strength required.

The copper comes from recycled brass and a giant builder's bag full of old taps and plumbing waits near the furnace to be melted down and transformed into a new life. A white metal finish is preferred for some items, such as spurs, for which a nickel alloy is used. This time another builder's bag awaits, full of coinage: Mexican pesos from the 1980s and no longer legal tender.

The company is closely associated with the Society of Master Saddlers and the Loriners' Livery Companies in the City of London (*see* page 11). Students on the saddle-making course come to the foundry to see a Bronze Age process which is fully integrated into a progressive company. The goods produced at the foundry today will still be top quality for their great-great-grandchildren.

An open frame showing newly cast buckles.

CHAPTER EIGHT

Historic and General Purpose Tack Rooms

The three fine tack rooms outlined here are sadly no longer in active use. Two are Georgian examples which demonstrate the richness of our heritage and make us think of how many other tack rooms have been lost. The third is an opulent Edwardian saddle room (built in the Georgian style) in perfect condition and on public display.

Chesters Estate, Scottish Borders

The Chesters estate and its Georgian manor house lie in the peaceful countryside on the banks of the River Teviot. Its rich farmland grows barley for the Scottish Borders Brewery, one of the estate's many enterprises along with fishing and glamping in shepherds' huts and yurts in the extensive grounds. A short walk from the house is a palatial stable block where the tack room has been preserved untouched for 29 years. The door was locked on the day of a tragic hunting accident which befell the owner, John Ogilvie. Shortly after it was reopened, the family kindly allowed me to visit the room; nothing had been disturbed since 1981.

I stood silently in the doorway taking in the sunlight filtering through the cobwebs, layers of old leaves on the floor and a mantle of dust over everything. The tack room had come to a sudden stop on a winter's day a generation ago. Evidence of its last busy morning was poignant: an open tin of saddle soap, a horse rug draped over the end of a saddle horse and a bucket for washing down the tack standing ready.

There was mildew on the bridles hanging on the wall and the rug and saddle linings had been ravaged by mice. Despite this, it was a pleasure to see such a well-appointed, thoughtfully designed room and it had an atmosphere of having been much enjoyed in its time. It was a fine big room, its panelled walls were lit by two sash windows and prize cards were pinned around the wooden saddle racks and bridle brackets. Everything was well ordered and a tack-cleaning

Chesters tack room opened for the first time after twenty-nine years. The room was locked on the day of an accident to the owner out hunting and never used again. Everything was left exactly as it was on the morning of the hunt in 1981.

The Georgian stable block at Chesters, designed by William Eliot and built of local red sandstone around 1790. Its courtyard, carriage bays, looseboxes, tack and feed rooms, offices and workshop are all now semi-derelict.

table and large old-fashioned radiator completed the impression of it being a comfortable world of its own. A bakelite wall telephone was connected to the house and a clipping machine powered by a hand-cranked motor had been there since about 1920. The looseboxes, each with glazed ceramic mangers, surrounding the courtyard have also been empty since 1981 but are still full of memories.

Winsley Hall, Shropshire

This spacious saddle room, which has served six generations of the Whitaker family, is presently dormant but still has hunting and polo tack on the walls. The room itself is cleverly sited in a very well laid out stable complex built of local Westbury brick. The doorway is under the cover of a substantial archway entrance to the yard. From here the head groom could monitor everyone who came and went, as well as being able to keep an eye on the stabled horses through windows on the courtyard side. A bricked area for the muck heap just through the yard kept it out of sight but conveniently adjacent to the walled garden beyond.

The long table was used as a base by visiting grooms and for shoot lunches for at least sixty years which continue to the present day. The history is in the fireplace bricks; a cast-iron range with an integral water tank on the side was there before the woodburning stove. The tank was filled in the morning, and by the afternoon when the hunt horses returned, the water was hot enough to mix into the bran mashes along with Epsom salts and perhaps a little boiled barley. It's easy to imagine this room in its heyday on a winter's evening with buckets of mash covered with linen cloths, a big pan of linseed boiling on the range and a tea kettle for the stable staff.

The saddle room at Winsley Hall, built in the late 18th century.

Manderston
Berwickshire, Scotland

Manderston is another historic estate with an extraordinarily grand saddle room. It is part of a stable complex cited by *Horse & Hound* magazine as 'probably the finest stabling in all the wide world'. It was designed and built of expensive ashlar for Sir James Miller by John Kinross in 1895.

The Manderston saddle room with polished rosewood doors. These saddles would have been used for hunting with the Northumberland and Berwickshire hunts, the latter claims to be the oldest in Scotland and dates back to at least 1740.

Entrance to the Georgian-style stable courtyards is through a splendid pedimented arch, decorated with hunting scenes and flanked by Doric columns. Within the courtyard are coach houses, looseboxes, stalls and a handsome saddle room for hunting, racing and carriage driving.

The stalls are finished with brass mangers and heel posts, each has the horse's name lettered in gold on marbled panels on the wall and the barrel ceiling and looseboxes are of teak. The stable yard is the last word in Edwardian luxury and worthy of Rock Sand, homebred winner of the 1903 English Triple Crown (the 2,000 Guineas Stakes, the Epsom Derby and the St. Leger races). Only one horse, Nijinsky, has won the Triple Crown since World War II.

The saddle room is surely the most splendidly furnished in the country with marble floor and rosewood ornate picture rail, panelled doors, cabinets and saddle racks. It is now on display to the public and its grandeur suggests that even in the 1900s it was a showpiece for visitors to the estate. Today, its owner, the Lord Palmer of Manderston, generously hosts the Berwickshire hunt at the house for a lawn meet as well as the annual hunt ball.

Ornate rosewood bridle bit cabinet at Manderston. A box of equine teeth ranging in age from foal to adult can just be seen, as well as the wooden handle of a balling gag (top) used to administer medicinal balls into the horse's mouth.

A magnificent table of Italian marble with brass legs stands over an original radiator in the centre of the saddle room.

GENERAL PURPOSE TACK ROOMS

Thousands of general purpose tack rooms are found throughout Great Britain, owned by families or belonging to livery yards, riding schools or trekking centres. There is a wonderful variety of old and new, grand and humble but all deserve to be appreciated and their contents cherished. Here are four examples of general purpose tack rooms, all greatly valued by their owners – and these are just the tip of the iceberg.

Speddyd Farm
Denbighshire, Wales

This tack room, which was once the bakery for the house, belongs to the Owen family who have farmed the Vale of Clwyd for generations. The brick-lined bread oven and Coalbrookdale solid fuel stove are still there, reminders of an earlier, self-sufficient age. During renovations to the building, a clay pipe was found hidden inside the wall, a local custom thought to be the 'signature' of the builder.

The room is now full of hunting and Pony Club tack, including the daughter's side saddle, top hat and gloves for equitation classes. Her grandfather, Phil, saddled his horse from this tack room and loved to ride along the Roman Road and Offa's Dyke in the Clwyd mountains when farming duties permitted. He continued to ride into his late seventies and enjoyed his hunting so much that he was buried 'proudly wearing scarlet' in his full hunting attire. His son and grandchildren share his passion and ride with the Flint & Denbigh hunt.

The oven, with its arched brickwork roof, is still intact, except for an ivy root that has worked its way through the stone wall from the outside. On the left is a hunting crop, a couple of leather girths and a twitch for holding a horse steady while dosing it with medicine.

The farm used heavy horses and a pony and trap until 1948 but none of this tack has survived. Periodically through the year the whole family would turn out on ponies to bring the sheep off the mountain for shearing, dipping and lambing. Quad bikes have generally taken the place of ponies but sheepdogs are still used. A few miles away along the mountain, Glyn Jones, thrice winner of the BBC television competition *One Man and his Dog*, used to train his young sheepdogs, many of which were exported all over the world.

Tack rooms are also useful spaces for other things – here the dog cage on the right has seen several litters of Jack Russell puppies and the side saddle above is safely out of harm's way. The best local potatoes from neighbour, Ronnie Evans, also keep cool in here.

In contrast to this vernacular tack room built of stone and with exposed rafters, a modern one has been built across the yard. This has all the latest facilities including a hot horse shower, heat lamps and integrated feed room and stables. Both tack rooms, old and new, have very distinct characters and are examples of the enormous range of tack rooms to be found throughout the country.

Cil Llwyn Farm Denbighshire, Wales

This tack room on a smallholding now houses a mixture of agricultural and riding equipment but it used to be occupied by a large Friesian bull. Inside, swallows nest in the rafters and until the 1950s farm carts and gigs were kept in open bays alongside. Press cuttings and rosettes line its lime washed walls and hemp ropes, English ploughing collars and breeching with heavy brass buckles have their place next to

The old brick Coalbrookedale stove (left) heated water for the bread and now makes a good shelf for hunting boots, a top hat and gloves for the side saddle equitation classes and the family's riding hats. The bread oven (centre) was used by the farmer's wife until the end of World War II and kept the room warm for all the hunting and agricultural tack stored here.

hunting saddles, bridles and foal slips. On the floor are chains for forestry work, farriery tools and a heap of old horseshoes, each one made to measure.

As a tribute to Red Rum, the Grand National legend, photographs of the Formby sands where he trained and of his statue at Aintree by Philip Blacker hang next to family horse portraits. A thick glider tow-rope is neatly coiled in a corner, waiting for next year's annual tug o' war between junior Pony Club members and Neville, a heavy horse. On a shelf there are plaiting needles stuck into a ball of wax, bottles of citronella to keep flies away, a jam jar of goose fat to protect against cracked heels in winter and grooming kits.

The room's contents span at least 150 years. This is a practical tack room, nothing smart but sufficient for the needs of hunting, riding and farm tasks – and a pleasure to be in.

Gelli Livery Yard
Flintshire, Wales

A livery yard is a lively social place with a variety of tack. Gelli is a small, family-run livery business catering for eight horses. The tack room and individual horse pens are all inside a modern, agricultural barn, which is covered in a canopy of pink *Clematis montana* flowers in spring. When not out grazing, the horses in the barn can lean over into the adjoining pen to scratch each other's necks with their teeth; a sign of trust and relaxation. Stabling design has begun to change away from traditional high walled looseboxes and to respect the horse as a herd animal. An increasing number of barns where horses are collectively stabled have looseboxes built with partitions low enough for the horses to be able to see each other and to make physical contact.

The tack room is tidy and organised so that each owner can keep their belongings separately in plastic boxes on shelves under a counter. Crumpet, the cat, has her base in the Adidas bag on the table and welcomes everyone, as does Hetty, a quiet and very friendly spaniel. Emma and her family who own the yard, designed and fitted out the tack room themselves. Large wooden racks, built by her father,

Behind this battered door, held shut with a nineteenth century iron pin about two feet long, is a good collection of riding saddlery and heavy horse harness. The pin was once used as an agricultural thermometer; thrust deep into a haystack it conducted heat from any fermentation in the centre. Whenever the farmhands passed by the stack, they would touch the pin to test its temperature as its warmth could warn of a serious fire risk.

keep the rugs tidy, the tiled floor is easy to clean and there is a propane heater for the winter.

Emma is responsible for clients' horses for the five weekdays of full livery and the owners take over at weekends. She will exercise or school the owners' horses in the manège, if required, and she also finds time to compete in one-day events at a national level.

The tack is very varied, as are the horses and their owners. Emma favours Amerigos saddles (Italian) and Neue Schule bits (German) made of an alloy rich in copper and nickel-free. This comparatively warm and 'soft' metal encourages

Nicky cleans a saddle while Emma tries to read with the cat and dog joining in. A wide variety of saddles and bridles are here, each owner of horses at livery in the yard having their own preferences.

saliva formation and is comfortable for many horses, as judged by their performance. She uses a Fulmer snaffle for dressage, a Tom Thumb sweet iron bit with copper roller rings for cross-country and a Universal, with several options for control, for show jumping. The rest of the tack is from a range of manufacturers and appropriate for dressage, hacking out and Riding Club activities.

There are lots of advantages to the livery system: plenty of opportunity to exchange information, frequent observation of all the horses when not being ridden and company when hacking out, if desired. The social side is illustrated by fond recollections from another livery owner who remembered that in the winter, everyone in the yard would congregate in the tack room to clean tack and sing along with the Top Ten on the radio, drowning out the drumming of the rain on the roof with their voices. The tack room was also a rendezvous for the occasional romance during the Christmas party.

Most of the owners work during the week so ride in the evenings and at local Riding Club events at weekends. These are times when the tack room is at its busiest and where everyone can chat over a cup of tea, catch up with equestrian magazines at the table or clean their tack.

It is also here where the affection for each individual horse is most evident. These are not just horses; they are treated as members of their owner's family. Given the expense and effort of having a horse at livery due to lack of time or facilities, these horses are also special companions.

This is an ideal way for working people to achieve a good work-life balance and to enjoy the ownership of a horse for competing or simply hacking out.

Mull Pony Trekking Centre Argyll and Bute, Scotland

This unusual tack room on the Isle of Mull, Inner Hebrides, is a flat-roofed wooden cabin. An adjacent wooden shed is the office and meeting place. Both buildings blend well into their surroundings and are

Dr W.R. Cook, a British equine vet, designed this bitless bridle so that the reins connect directly onto the noseband through a ring, crossing under the horse's jaw.

The tack room for the Isle of Mull Pony Trekking Centre in Killiechronan. This discreet wooden cabin houses all the tack and equipment needed for some of the most exciting trekking in Britain.

an environmentally sound and practical choice for a summer trekking centre in remote and beautiful countryside. There is no electricity or phone line: mobile phones and long summer days make both unnecessary. There is also no need for horse rugs and drying facilities as these hardy Highland ponies know where to find natural shelter all the year round in their extensive grazing. They are used to the wind and high rainfall and the tack here has to be able to withstand the wet.

Bitless bridles and Australian stock saddles are sometimes used for less experienced riders but otherwise the tack is varied. Apart from trekking visitors, local riders also enjoy learning here, competing on the trekking ponies at the Killiechronan annual show and taking part in Pony Club activities. A Mull branch of the Pony Club was established in 1976 and runs a full programme of events with summer camp, hunter trials and instruction. Head collars of nylon webbing and saddles and bridles made from synthetic leather look authentic and, unlike real leather, dry quickly after riding in the sea or the rain. The owner, Liz Henderson, finds that bitless Dr Cook bridles (which can double as a head collar when the reins are detached) are the most suitable for the trekking clients. These bridles are effective in controlling a keen pony through pressure on the nose, poll and under the chin. They must be used very sympathetically but have the advantage in a situation where riders may not be very experienced, of protecting the ponies' mouths from misuse.

Because of the steep terrain, the Australian stock saddles are ideal as they help to keep novice riders in position going up or down hills. Trekking here in the wild and natural scenery of the West coast of Scotland is an artistic as well as an equestrian experience, particularly in October.

In one day the weather can be a kaleidoscope of moods; a calm morning, the peat smoke drifting among the autumn leaves and bright shafts of sunshine on the sea, can suddenly turn wild and dark under huge woollen clouds. A Westerly races in off the Atlantic. The gale whips away the lacy streams flying down rock faces creating fans of faint rainbows in the air. It roars away over the moors and heavy rain follows, easing into quiet veils of mist as the afternoon cools towards dusk.

I joined a group of four other visitors and we were each allocated a pure or part-bred Highland pony. Our three-hour ride began with a lively canter along the beach. Then with no warning, a hail storm burst out of a clear blue sky, rattling on our helmets

Trekking along the beach on the Scottish island of Mull.

and wax jackets. In one practised movement, the ponies stopped in their tracks and spun round with tails buttressed against the stinging ice. With all heads down, hunched against the storm, it was only ten minutes before blue sky returned. We forded a river deep enough for the water to reach our feet but soon dried out during a steep climb of three hundred metres straight to the top of the headland, standing in our stirrups as the ponies scrambled up. After a lengthy pause to admire the panorama, we turned inland and ambled gently downhill on a loose rein through heather and bracken back to base.

On the previous day up on the same headland, the riders had been treated to a rare sight: a sea eagle, a raven and a fish eagle all feeding together on a sheep carcase just a few yards away. I saw a buzzard, statue-like, on a post nearby, wild geese resting in a field alongside the single track road and several deer in the distance.

The native ponies are such a natural part of the environment that the wildlife seemed to be entirely at ease as we passed. After the ride, the wet, salty and sandy tack was simply wiped over with a damp cloth and hung up, ready for the following day.

CHAPTER NINE

Beach Donkeys

The donkey originated in the Asian deserts yet its image goes straight to the British heart. Is it because it reminds us of Jerusalem with the marking of the cross on its shoulder or its bizarre braying and extraordinary ears ridiculed by Shakespeare? Or is it because of its long-suffering nature immortalised in Modestine by Robert Louis Stevenson in his *Travels with a Donkey* in 1879? For whatever reason, no British seaside holiday would be complete without beach donkeys. Their huge ears, exotic eyes and gentle nature make them a favourite with everyone.

In coastal resorts, such as Weston-super-Mare, donkeys have a long history. They were first used for hauling the fishermen's catch onto the shore and later as draught animals delivering milk and coal. From 1840 they also provided a taxi service pulling donkey chairs round the town and giving rides on the beach.

Beach holidays have been recorded in diaries from the early 1820s when the gentry travelled to the seaside for their health. A huge rise in popularity followed after World War II when special trains such as *The Merchant Venturer* carried factory workers from the major industrial cities to the seaside for their annual holidays. The Great Western Railway produced posters of sunshine and happiness at coastal resorts which are now collectors' items. The 1950s were the heyday of the Donkey Derby races which were often supported by celebrities such as Charlie Chaplin and Laurel and Hardy and countless images of laughing 'jockeys' must have been recorded on Brownie cameras.

Today's donkeys at Weston-super-Mare are of Irish, British, French and Spanish stock and come in various sizes and colours. A good-sized female donkey (a jenny) can be worth about a thousand pounds.

Terry Vincent of Weston-super-Mare, a donkey owner, first helped with the beach rides when he was eight years old, some 55 years ago. His family, the Trapnells, have owned and bred donkeys since the early 19th century.

In days gone by they were herded daily through the town to the beach, but now they would cause such disruption to the traffic that they are driven by horsebox from their sixty acre farm nearby. A fifteen ton Mercedes horsebox doubles as their tack room. It is parked on the sand and gives shade to the donkeys in the heat and shelter from the wind. Inside, the space over the driver's cab is filled with saddles and bridles. The salty sea air blows in through the open slats high up on the lorry sides encouraging rust to form on the stirrups and stiffening the leather. It is an essentially practical tack room with equipment

William Parsons of Weston-super-Mare with his donkey chair, c.1890.

Donkeys on Weston-super-Mare beach. None of the riders appear to be holding the reins but the donkeys know their job and keep within their territory.

and donkeys always together in the same place and it works well.

Donkey saddles and bridles are adapted from pony tack to suit their distinctive anatomy. The browband, stamped with the donkey's name, is lengthened to fit the broad brow and two girths are needed to stabilise the saddle as donkeys have no pronounced withers. The second girth is attached by splitting open a pony saddle and screwing the girth strap onto the tree so that it fits further back around the belly. A metal handle is welded over the pommel to the stirrup bars on either side of the saddle to reassure nervous riders.

The method of selection of individual donkeys to go to the beach for the day is interesting and, according to a Blackpool donkey man, the donkeys made their own choice.

'Early in the morning, the lorry was driven into the field where some forty donkeys grazed and parked with the ramp down', he said. 'While the driver went to have a cup of coffee, the donkeys who wanted to go to the beach, loaded themselves. After several days working at the beach, those donkeys would decide to stay in the field for a while and others took their place. Generally, about a dozen donkeys would choose to go to work each day on the beach.' Like most donkey owners, he had great respect and affection for his animals.

Spare donkey saddles in the horsebox 'tack room' showing the wear and tear of beach work in the salty sea air. They are characterised by a second rear girth and a metal handle attached to the pommel.

Young donkeys, four years old, are gradually introduced to the challenges of beach life which include frisbees, kites and musical carousels.

Donkeys are not broken in for riding in the same way that ponies are. After accepting the snaffle bit and saddle, the animal is first ridden by a competent child while in the company of the herd on the beach and then learns its trade by following the example of experienced donkeys, the process normally only taking a few days. The donkeys are unshod and each is identified by a number on its hoof just below the coronet.

Today's high welfare standards ensure that working donkeys lead a contented and comfortable life, many spending their retirement in a donkey sanctuary. The donkeys work a maximum of a six-day week, depending on their age and experience, and have access to hay and water throughout the day. No child older than fourteen, or over eight stone (50.8 kg), is permitted to ride a donkey. During the winter the donkeys go on holiday, often billeted with families who live nearby.

Although there are still several donkey owners working on the beach at Weston, the number of donkeys has declined along with the fortunes of the British holiday resorts and the necessary skills for their management are becoming harder to find in the younger generation.

The short peak season is from the end of May to September and the weather is unpredictable. This

Terry Vincent's donkeys in Weston-super-Mare enjoying ad lib hay on the beach.

means that Terry and his business partner, Richard Warburton, must be on the beach daily with the donkeys to offer rides, whatever the weather, if the business is going to prosper.

In addition to the donkeys – and to cater for older children and adults – two handsome cobs, Sid and Appleby, harnessed with collars and breeching, pull smart wagonettes at a spanking trot along the wet sand. With the sea breeze blowing in their manes and the sea sparkling in the sun this must be the best value for £1.50 in the country.

Different beaches in Britain have their own local rules. Blackpool beach has a Blue Flag award which means that it complies with the international standards of safety and hygiene agreed in 2001. This is good for tourists but the donkey owners have to ensure that all the droppings are removed. Blackpool city council demolished the original donkey stables and tack room in the 1970s and now issues licences to

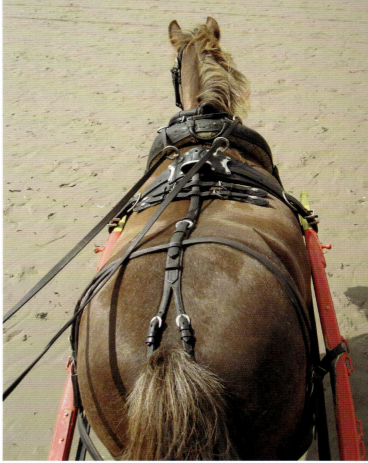

Appleby pulling a wagonette full of holiday-makers at a brisk trot towards the sea.

private owners. Despite the constraints, donkey owners, some of them fourth generation in this beach work, organise a rota for the best pitches on the beach according to the tides and still provide very special traditional entertainment.

Essentially beach life has hardly changed over the last hundred years. Stoic and intelligent, the donkeys wait untethered and dozing in a group. They seem to enjoy the noise of family beach ball games, the Punch and Judy shows, candy floss and ice cream stalls. The sound of the gulls, children laughing, the jingling of the donkeys' harness and the waves on the shore is still a quintessential British experience.

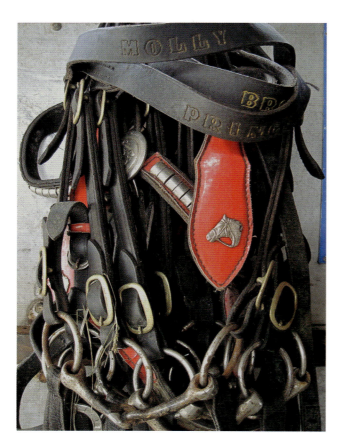

Donkey bridles with extra broad browbands. The brass buckles do not corrode in the sea air, unlike the bits.

CHAPTER TEN

The Wadworth Brewery

The Wadworth brewery in Devizes is one of the few remaining breweries which still uses heavy horses. It is the centre of 'Wadworthshire', an area of Wiltshire and surrounding counties which includes most of the brewery's two hundred and fifty or so tenanted pubs and hotels.

Appropriately, this is White Horse country where eight renowned folk-art horses grace surrounding hills, their outlines cut out of the turf to expose white chalk. The oldest (Uffington) is three thousand years old and the most recent (Devizes) was created in the year 2000 to celebrate the Millennium. The Late Bronze Age Uffington horse is likely to have been made in honour of the Celtic goddess Epona, patron saint of the horse. She is often depicted in bas reliefs protecting horses and foals decorated with ears of corn.

Horses and corn are still fundamental to Wadworth's Brewery today: horses for the delivery of the beer brewed from barley. It was founded in 1875 and for the last hundred years their famous Shire horses have been used for local deliveries in Devizes.

This family-run business upholds traditional crafts and employs in-house sign writers for producing those uniquely British pub signs. No computers are used in the art work; it is all drawn freehand and real gold leaf shines through all weathers. Until recently, the last brewery Master Cooper in Britain also worked here but the introduction of metal kegs instead of oak barrels to comply with new hygiene laws threatens the future of cooperage skills.

The Wadworth Shires, currently Sam, Max and Archie, are dark bay with four white stockings. Each weighs just under a ton (at least 800 kg) and Max stands 18 hands high. Traditionally each brewery prefers horses of a particular colour and matched within a team. Five days a week these horses head towards Devizes pubs and shops, unfazed by traffic, and the local constabulary ensures squad cars mute their sirens when the horses come into view.

When I visited the brewery, Prince and Monty, the two Shires employed then, had just been harnessed to the loaded two ton dray for the morning beer delivery. Head Horseman, Martin Whittle and Horseman Barry Petherick (now succeeded by Callum Whittle) wearing traditional flat caps, climbed up onto the driving seat and the horses pulled out eagerly, Monty trotting on the spot, before both settled into a majestic clopping rhythm which echoed down the Georgian street.

Perhaps invigorated by the scent of hops and malt on the breeze, they obviously enjoyed their work and no doubt had their own agenda too; admirers were waiting along the delivery route with treats for the horses. The dray pulled up outside the first pub, the wheels were chocked and a waterproof half blanket thrown over their backs. This is what they had been waiting for! The faces of the fan club of old ladies and children lit up with pleasure as the huge, whiskery muzzles delicately accepted their offerings. The chopped apples and carrots held out on the smaller palms disappeared in a peal of giggles. A

Monty and Prince at the brewery, loading up for the morning delivery in Devizes.

frail old man clearly derived as much pleasure from sharing his flapjack with the horses as they did before it was time for them to move on.

The enormous power and presence of the team never fails to stop shoppers in their tracks and lend a sense of occasion to the morning. They are a touchstone that all is well in Devizes. The horses are well integrated into the community and visit local care homes where their beneficial aura is appreciated by the residents.

On occasions they give wagon rides to the public and the horses' annual two-week holiday is an event filmed by TV. Crowds of over a thousand people go to watch the horses have their swig of Wadworth's most famous brand, 6X ale, before being turned out to pasture for their summer rest.

Wadworth is now the only brewery in England to own, work and show its horses. The brewery supports several shows every year, competing successfully in turnout classes – a very demanding schedule as horses, harness and handlers must be nothing short of immaculate. A full turnout should include lamps, a wooden bucket, nosebags and blankets for the horses while the driver has a bowler hat, whip and protective knee rug.

A summer dawn on a show morning will see the horses having their white feathers washed and powdered with woodflour, the raffia flights (blue for Wadworth) being threaded into crest of mane and top of tail, coats groomed and hoofs oiled. The show harness is carefully wrapped and loaded with the dray and the horses into the twelve-wheeled articulated Mercedes trailer which also has living space for four people. A show is an occasion to fly the Wadworth flag and to demonstrate the skills of driving a pair or four in hand. It is also a chance to socialise with the public and provides a break in routine for the horses. The presentation of these magnificent teams is the culmination of a lifetime of experience in training and caring for these working horses, keeping them fit and happy, safe, and leading productive lives.

Behind the scenes, in the stable yard, a steady routine is the key to success. The horses are stabled next to the feed store where bran and brewer's grains are kept in oak casks handmade by Wadworth's Master Cooper, Alastair Simms.

After an early morning feed the horses are mucked out and groomed, ready for work at 8.30am. They return for a lunchtime feed and rest for the afternoon when members of the public may visit, or the farrier may attend to them. A set of shoes lasts only three or four weeks due to the weight of the horse and daily roadwork. Two more feeds and a late night check-up by the staff complete the day.

The horses are bedded down on peat and mucking out, the mythological fifth labour of Hercules, is well organised here. Peat has the advantage of being a good insulator, soft, absorbent and most importantly, doesn't harbour the pasture mites common in straw which cause itching and sores in the feathers of heavy horses.

Unusually, this stable yard doesn't have a muck heap. Peat and droppings removed from the stables are put into bags and sold to the public for garden compost.

Alongside the stables and in the original stone buildings is the Harness Room, the hub of the yard. This is a light, bright room warmed by overhead heating pipes. Pine panelling keeps the harness from marking the walls, and wide benches with cupboards below are arranged in a T-shape in the middle of the room. Every afternoon two or three staff clean the harness, chatting away to visitors and putting the world to rights. You can see your face in the brass name plates below each collar and admire the dozens of show rosettes pinned round the top of the panelling.

The wheeler and leader sets hang next to the single horse harness with its cart saddle.

Bridles for showing in-hand, a set of red swingle trees and the daily working harness line the rest of the walls. Most of the harness is made by Master Collar Maker Terry Davis in Ludlow, using bridle and strap leather 5mm thick and oak bark tanned by J. & F.J. Baker in Devon. The show collars are leather

The harness room where Martin Whittle and Barry Petherick clean the working harness after every morning delivery.

lined and the outer part (the aftwale) is made of black patent leather for extra dash and sparkle to catch the judge's eye. The fixing chains are shiny chromed steel and horse brasses with the Wadworth motif of a cask within a laurel wreath decorate the bridles.

The English working collars are stuffed with rye straw and lined with traditional woollen collarcheck which is absorbent and hardwearing. The quality hames are double-cased in brass rolled around a wood core and stamped with the maker's mark. Liverpool bits are used throughout; those used in pair driving have the shank reduced to prevent catching in the companion horse's bridle.

This draught harness has evolved through trial and error over hundreds of years to reach its present design and is excellent for the horse's comfort, safety and working efficiency.

All these traditions at Wadworth are an example of sustainable work practices which also give enormous pleasure to the public. Long may Epona's vigil over the horses continue.

Some of Wadworth Shires' show harness and prizes.

Happy holidays! Max, Monty and Prince at the start of their annual fortnight's holiday.

CHAPTER ELEVEN

Canal Horses

Horse-drawn boats and barges on the canals, the arteries of the Industrial Revolution, were vitally important for about one hundred and fifty years beginning in the 1760s. Tens of thousands of horses, as well as mules and donkeys, hauled boats carrying raw materials and fragile products in quantities which could never have been supplied by road.

A packhorse could carry about two hundredweight in its panniers, a cart horse could pull up to one ton in a wagon but a canal horse could pull a barge of about eighty tons. As the boat horses drove the pack horses out of business so they themselves were eventually overtaken on some canals by steam driven tugs which could tow a string of barges, then by the railways and finally by inboard diesel engines on the boats.

The railways became increasingly serious competitors from the mid-1800s though horse-drawn boats continued on the Leeds & Liverpool canal and in the Midlands until 1955 and horse-drawn barges on the Regent's Canal, London, until the 1960s. Little remains to remind us of this history. Most telling are the deep grooves in hardwood posts and iron plates on the corners of buildings to protect them from abrasion caused by grit in the horses' tow ropes wherever the canal path changed direction.

Boatmen, their families and horses were a unique community, fairly isolated from both urban and settled rural populations. This specialised life, physically tough and materially frugal had its own jargon, traditions and rules. The canal horses also became adapted to their task and were said to see out their days on the towpath, rather than being sold on for farm or market garden work. Accustomed to pulling loads at an angle, the farmers found that these horses would often have their forefeet in one row of crops and their hind feet in the next, so they were not popular. They were also reluctant to reverse as they had never been taught to push back into breeching. Old photographs show that the boat horses wore traditional trace or plough harness (*see* Harness page 34) often called 'gears' which was adapted for the job. The bridle normally had blinkers and a basket muzzle could be attached to it to stop the horse from eating the hedge while it worked. In the summer, white ear protectors crocheted by canal women were put on to prevent irritation from flies.

The English collars had hames cut down to just above the top of the collar to avoid catching on bridges and the traces, often of rope rather than chain, were threaded through wooden bobbins to prevent chafing.

The towline was attached to a whipple tree (a stretcher or swingle tree, depending on the region) which was held above the hocks by brass buckled quarter straps. Hames, bobbins and whipple tree were usually painted in primary colours or in the livery of the owning company. Sometimes a shoe or pan was tied by a long string to the whipple tree to fool the horse into thinking it was still being driven when the driver had hopped on board for a rest.

All types of horses were found working on the canals, large heavy horses were needed for barges but cobs up to 15hh could pass more easily under the low bridges on the smaller canals. Apart from hauling, others worked in construction and maintenance

The Roundhouse, Ladywood, Birmingham, built in 1874. This striking horseshoe-shaped stable yard with its granite setts, elegant windows, haylofts and stalls is full of the ghosts of hundreds of canal horses. Two octagonal gatehouses flank the entrance on the land side from where the managers could monitor all movement.

by fetching and carrying materials, or on a 'gin' to pump out water from lock pits by walking in circles, blindfolded to prevent dizziness. While oxen were important agricultural draught animals, they were not as suited to the towpath as were horses. However, herds of cattle were driven up and down the newly dug canal to tread down the clay lining during construction, a technique known as 'puddling'.

The first canal constructed independently from any existing river courses was the Bridgewater Canal, owned by twenty-one year old Francis Egerton, Duke of Bridgewater and designed by James Brindley (1716-1772), a millwright with no formal education and a brilliant intuitive engineer. Its first section was opened in 1761 for the transport of coal from the Duke's collieries in Worsley to Manchester to power steam engines for the Industrial Revolution. He also started a stud to breed mules to haul the boats. Mules are strong, cheaper to feed and have harder feet than horses though they (and donkeys) never became generally popular on the canals, except on the Kennet and Avon canal where they are mentioned in folk tales and songs.

Stone steps were built into the bank at intervals, as on the Bridgewater canal, to enable animals that had fallen in to climb out. They could fall into the canal when passing one another on a slippery towpath or when pulled in backwards by the towline if they were overrun by the boat, its momentum being far stronger when moving than the horse could withstand. In 1851 Queen Victoria and Prince Albert stayed with the Duke of Bridgewater at Worsley Hall and were taken on a barge elaborately fitted out for the occasion. The horse steps came in handy when the horses shied at the excited crowds and fell into the canal.

Welfare Acts in the 1830s did much to improve the life of the boat horse and inspectors had the power to issue fines or dismissal notices for maltreatment. It was a hard life for a horse nonetheless. An average distance for a horse would be about twenty miles a day of hauling, eating from a nose tin strapped to its head while it walked. In the eastern counties where private land came down to the water's edge, there were stiles, about two feet six inches high, set into the fence across the towpath which the horse had to jump. A young lad, a 'horse knocker', would mount the horse, check that the towline was slack and jump it over the stile – as painted by John Constable in *The Leaping Horse*.

Another manoeuvre required from the horse, when there were no roving bridges available, was to scramble onto the boat when the towpath ended to be ferried across to the path on the other bank.

Tunnels rarely had a towpath for economic reasons and the boat horses were walked over the top to rejoin the boat when it emerged. The Standedge Tunnel (1811) in West Yorkshire is Britain's deepest, highest and longest tunnel (over three miles) under Marsden Moor. While the boat was legged through by men lying on their backs with their feet 'walking' on the rock roof and sides, the horse was led for four miles over the moor along the packhorse route. It was a hard life for canal workers too. In 1811, Thomas Bourne, aged twelve, was appointed as the Standedge traffic regulator. For thirty-seven years he led horses over the top four times a day, seven days a week and checked that the horse boats travelled safely. It is estimated that he must have walked 215,812 miles in all weathers during his life.

Fly boats for passengers or perishable cargoes were pulled by two or more horses at a canter, in tandem and with one postilion per pair. It is said that they travelled at about twelve miles per hour

Horseman Dave Poxon (see page 69) working in the Tiverton Canal Co. tack room. Behind him are two American collars, one with a vinyl underpad. A cart saddle and breeching are there for winter cart work to keep the horses fit in the off season. The brightly painted wooden trace bobbins and curved whipple tree in the corner are particular to canal harness. The use of anti-chafing bobbins was recommended in Acts of Parliament concerned with animal welfare in the 1830s.

Taffy pulling the passenger barge 'The Tivertonian' owned by the Tiverton Canal Co. on the Grand Western Canal in Devon. This photo, taken by Sarah Williams, won the 2012 BBC Countryfile Calendar competition which raised over one million pounds for Children in Need.

with a two minute changeover for fresh horses every four miles, though this would have been difficult to maintain in practice due to the condition of the majority of towpaths.

Canal transport in the winter was preferable to travelling in a coach along mired roads and specially designed ice-breaker boats with reinforced bows kept a channel open. Ice-breaking was an occasion when the canal community all came together. As many as a dozen horses were hitched together to drag the ice-breaker through. Alternatively, they pulled a boat with a rounded hull filled with men whose weight prevented the bow from riding up over the ice and which could be rocked from side to side to widen the channel.

Canal horses were stabled at night, both because a hot, tired horse left in the open might fall ill and because of regulations enforced during the 19th century.

Horses belonging to canal or railway companies were stabled at the local company depots where foremen and vets were on hand. Other horses, independently owned, would be housed in the stalls of an inn or farm building where conditions were variable. If no tack room was available, harness would have been hung on a peg opposite each horse's stall, easily accessible for the next shift.

In dubious places, 'confetti' could be added to their feed to prevent others from stealing it for their own horses – the tiny pieces of coloured paper were harmless but instantly identifiable. Up and down the country, a few remaining canal stables can still be found, often a simple four stall building owned by a farmer who made a small overnight charge. The walls would be whitewashed with slaked lime, its alkalinity acting as an antiseptic. Each stall had a wood block tied to the end of the horse's head rope which moved up and down through a fixed ring allowing some freedom of movement without tangling.

Birmingham claims to have more miles of canals than Venice and was a major transport centre in the 18th and 19th century. The Grade II listed Roundhouse stabling (1874) in Ladywood, is a monument to the importance of horse-drawn traffic on the canal. This unique horseshoe-shaped building was designed by W.H.Ward who won the Birmingham Corporation's architectural competition. It was built on top of a natural rise in the ground with stabling and haylofts for scores of horses.

A wide tunnel led under the building to the coal and mineral wharf and in the undercroft beneath the stables were workshops for harness repairers, feed merchants' stores and a farrier's smithy. The elevation of the building enabled all the barrow loads of muck from the horse stalls on the ground floor to run downhill to be taken away by barge, as well as providing maximum ventilation for the stalls. Horse manure was valuable and was mixed with hot tar and cow-hair to make 'chalico' which was used by boatyards as a dressing for the timbers of wooden vessels. Canal companies also sold manure to market gardeners and harvested the hay on canal banks for their horses' winter provisions.

When I visited the Roundhouse, before its transformation to a cultural centre with workshops, it had been derelict for decades. Little was left of the original stalls and it was impossible to locate the tack room. The industrial stables of the same period at Camden Goods Station on the Regent's Canal in London, though on a larger scale, certainly had a tack room which was presided over by the Horse Master.

The legacy from the days of busy commerce is now in the hands of just four horse boat companies where the same skill and judgement is needed to operate horse-drawn boats and barges, but the cargo today is passengers. At the time of writing they are located in Tiverton (Devon), Godalming (Surrey), Newbury (Berkshire) and Llangollen (Denbighshire).

Tiverton Canal Co. Horse Drawn Barge, Tiverton, Devon

The Tiverton Canal Co. is a family-run business on the Grand Western Canal, a stretch of waterway once part of an ambitious 19th century plan for a navigable route from the Bristol Channel to the English Channel to avoid the long hazardous journey round Land's End. Devon County Council now owns the Grand Western Canal, completed in 1814, important then for bringing quarried lime to the area for agricultural use and coal. The canal today is part of a country park and local nature reserve.

Philip Brind started work at the horse-drawn barge attraction back in 1986 and with his wife, Jackie, bought the Company from his parents in 2005. The staff, who include daughter Becky and son-in-law Dave Poxon as horseman, wear old-fashioned costumes, and create a real feeling of going back in time. Their horse-drawn passenger barge, The Tivertonian, weighs in at about thirteen tons, has a ten foot beam, and her seventy foot length is brightly painted with the traditional folk art of roses and castles.

Having started the job about thirty years ago, Phil knows every mile of this canal in all its moods, the different echoes under each bridge and where the moorhens nest. He is proud to run one of the last horse-drawn barge businesses in Great Britain and understands how important it is to keep this living heritage alive for future generations. Heavy horses are built for work and Phil takes much satisfaction in being able to provide them with a job they are suited for and, from their attitude, clearly enjoy.

A barge horse learns the skill of towing, has plenty of variety, keeps fit in the fresh air and gives pleasure to the many visitors both local and national.

I visited on a sunny May morning and opened the gate into a grass paddock on the canal bank with views across the Devon countryside. The paddock was bordered by a row of stalls adjacent to a tack room and George, the novice barge horse, an eight-year-old part-Shire, was waiting there to be harnessed.

Although already broken in for riding and driving, this was only his second lesson in pulling the barge.

George was harnessed in his familiar bridle with blinkers and an American collar with hames and vinyl underpad which clips to the forewale of the collar. Underpads are usually made of cloth stuffed with wadding but these can take days to dry and wet collars cause sore shoulders. The vinyl pad, filled with a layer of foam, is comfortable in all weathers on the horse's skin and easy to clean. The Company has several horses and American collars can be adjusted through 21-24" to fit different neck sizes.

The rest of his harness was traditional, as described above, with traces threaded through bobbins and running back to the whipple tree to which the rope towline is attached.

Out on the towpath, George and The Tivertonian were hitched together and Dave took up the long leather reins attached to the Liverpool bit to guide the horse from behind. Phil was at George's head and I walked alongside.

Pulling a barge is very different from everything George had previously experienced. A cart moves off instantly when the horse goes forward, but a stationary barge has inertia and needs a good strong pull to get it moving through the water.

Once away from the mooring it soon gathers momentum and keeping a steady tension on the tow line requires little effort, in fact the barge can be moved by a single man. With encouragement from Phil and Dave, George found the confidence to keep pulling until the barge moved and then kept up a brisk walk, alert and with ears pricked. George has a sociable and willing nature and was making very good progress. Unlike many horses he will be meeting dozens of people every day and sharing the

A novice horse, George, learning to tow the passenger barge (out of the picture to the right). The towline runs from the luby pin on the barge's towing mast to the hook on the whipple tree. The line of draught is straight from the mast along the traces to the hame hooks on the collar, the most efficient way to pull a load (see Principles of Draught page 38).

towpath with walkers, bikes, pushchairs and dogs.

Both Philip and Dave were in communication with the tillerman on the barge. As there is no engine, they constantly monitor the angle of the boat and position of the horse, particularly when navigating bends in the canal, the tension on the towline and the momentum of the barge. It looked deceptively easy but is a skill only acquired by considerable experience.

George had already adapted to the asymmetric pull from the hundred foot long rope towline as well as to the rustling sound it made behind him as it passed over the yellow irises and cow parsley growing along the banks. The steady rhythm of his walk caused the line to slacken and tighten gently, though it was barely noticeable. The whipple tree and traces moved slightly up and down on George's flanks and the anti-chafing action of the bobbins was immediately clear as they rolled smoothly against his side.

Leaving the last few Tiverton houses behind, the canal runs out into the countryside and we passed a wildflower hay meadow which will be cut for the Company horses' winter provisions. Compared with a road, this waterway was quiet except for George's soft footfalls on the towpath and the birdsong. There were no diesel fumes, only the scent of blossom. A kingfisher flashed by and the trees were mirrored in the still water surface. This slower pace gave us time to reflect on the heritage of the extraordinary engineers, such as Brindley, Jessop and Telford of the 18th century.

This canal once recharged the local economy and now the tranquillity of its wildlife recharges the many naturalists, anglers, walkers, cyclists and horse-boating public who share the park. The horses too, adept in predicting the motion of the barge, clearly enjoy their task and are perfectly suited to the work.

Canals were nationalised as British Waterways in 1948 and since 2012 most have been managed by the Canal & River Trust with Prince Charles as its patron, though ownership today varies.

The Horseboating Society was founded in 2001 at the National Waterways Museum in Ellesmere Port, Cheshire to promote the skills of horse-boating wherever possible on the national canal network.

It has celebrated several bicentenary celebrations including that of the Standedge Tunnel. On April 4th 2011, the boat horse pulled 'Maria' to the Diggle end of the tunnel where it unhitched. The horse 'Bilbo' was led over the top to the tunnel exit at Marsden while the crew of six 'legged' the boat through by the light of a Tilley lamp. After three hours they emerged tired and dirty but elated. The waiting silver band struck up and after speeches and applause we all headed for the pub.

Glossary for Canal Horses

Whipple Tree or **Swingle Tree** – *see* Harness page 34

Gin: A horse-, donkey- or mule-powered machine for removing water (or other materials) from below ground level. A long horizontal arm was attached to a central wheel which engaged with gears going down into the pit to bring up buckets of water. The animal was hitched to the end of the arm turning it by constantly walking in a circle.

Roving bridge: A bridge over a canal constructed with a horse walkway which enables the horse to leave the towpath on one side and join the path on the opposite side.

Canal boats and barges: An accepted definition for a canal boat is a **narrowboat** with a beam of less than seven feet which usually provides living accommodation.
A **barge** normally describes a wide beam vessel of about ten feet which carries a cargo.

Luby Pin: A quick-release iron pin on top of the towing mast for the attachment of the towing line. The mast is positioned on the central axis of the boat at a distance from the bow which is neutral (no bias) to the forward towing force. This means that when towed forwards, neither the bow nor the stern tends to move towards the bank.

CHAPTER TWELVE

Household Cavalry Mounted Regiment and King's Troop Royal Horse Artillery

The skills of ancient and modern warfare combine in the Household Cavalry and the King's Troop: both form the personal mounted bodyguard of the Sovereign and are also fighting units of the British Army on active service.

At the daily inspection parade of the Household Cavalry, immaculate black horses stand to attention, their riders' helmets shining and plumes waving in the breeze as the pomp and ceremony begins with the rallying call of the trumpet.

Every morning since 1660 this cavalcade has been on duty to protect the monarch and guard the entrances to the Royal Palaces in London. Yet the soldiers on these chargers bridge the gap between the cavalry era and current conflicts with an extraordinary mastery of both horsemanship and technology. Their dual role is to carry out ceremonial duties on state and royal occasions and to have operational roles in reconnaissance using armoured fighting vehicles in theatres of war. They are presently on duty in Afghanistan and Iraq.

The Household Division is made up of seven regiments, comprising the Household Cavalry (The Life Guards and The Blues and Royals) and five regiments of Foot Guards (Grenadier, Coldstream, Scots, Irish and Welsh).

The Household Cavalry is the most senior regiment in the British Army and traces its origins back to Charles II in the 1660s. Their distinguished campaign histories include Waterloo, the Peninsula, Crimean and Boer Wars, Ypres, the Falklands and Afghanistan among others. Their ceremonies of Changing the Guard, Trooping the Colour on the Queen's birthday and their Musical Ride are internationally famous.

The Household Division at the Hyde Park Barracks (Knightsbridge) is a stone's throw from Harrods department store, and occupies stabling on two floors which was designed by Sir Basil Spence in 1970. Over two hundred horses can be accommodated together with the teams of saddlers, farriers, tailors, vets and riding staff who make this a versatile and self-sufficient force. On entering the Barracks one is immediately aware of the transition from noisy chaotic London to a highly ordered world. The bonds of trust and respect between every soldier are apparent; there is an enviable confidence in their knowledge that in a conflict crisis this fellowship will not fail. In this special community the privilege of being a member is clearly a matter of great pride and honour.

The tack is divided between two areas: one for the civilians who help to exercise the horses in Hyde Park and another for the ceremonial occasions. The first houses a mixture of contemporary bridles, bits and saddles and the heavy-duty waterproof capes worn by the riders.

The Household Cavalry returns to Buckingham Palace after the Trooping of the Colour to honour the Sovereign's official birthday on 12th June.

The second is a large, windowless room where the 'horse furniture' used for official duties of the Household Division is kept. A huge central table with drawers to the floor fills most of the space. Around the outside are the saddle and bridle racks and along the length of one wall are humidity-controlled glass cases which protect the ceremonial bridles used in the presence of the Queen. The brightness of the lights catches the elaborate metalwork of the regimental capstars on the bridles and the jewel colours of scarlet and gold embroidered saddle cloths on the table.

The tack is just as it was when the horses were last fully engaged as cavalry up to the end of the First World War. Only the saddles have gradually been modernised and made more comfortable for both mount and rider. The general purpose cavalry saddle, developed from the original Hungarian light cavalry design, was modified in 1902 and 1912 to become the British Army Universal Pattern. Every horse has a custom-fitted saddle and bridle with his or her name stamped on the leather which is made by the resident master saddler and his team.

Recently a new Pathfinder saddle with extra panels of padding which help to distribute the weight

Household Cavalry officer during morning inspection at Hyde Park Barracks. New pairs of jackboots are covered in beeswax which is melted using a blowtorch and polished to establish a high shine which improves with age and elbow grease.

more efficiently over the horse's back has been tested. Each soldier carries about four stone of kit and horses must stand patiently, often for long periods of time.

Honed from long experience on gruelling campaigns, every piece of tack had a function, almost all of them having been designed for use in battle and they are still retained today. Nor was much wasted: for example, in the First World War, horseshoes were made from melted down German guns and nails from French bullets. The war economy affected non-military horses remaining in England too and strict rationing rules for their diets were pinned up in every tack room in the land.

A chain neck strap connected to the back of the noseband could be used as emergency reins and the horsehair plume hanging from the throat lash of a Blues and Royals officer's charger helped to protect its throat from an upward sword thrust. The substantial brasswork on the bridle headpiece is not only decorative but protected the horse's vulnerable poll during battle.

The crossbelts worn by officers served as spare girths and their bearskins or astrakhans on the saddle could be used as a cover for nights in the open. For ceremonial occasions where a member of the royal family is present, shabraques (saddle cloths with embroidered borders) are used instead of the bearskins. A major-general's kit could include a leather rifle bucket and pouches strapped to the saddle which held pistols, maps and writing materials. Gold aiguillettes (braided cord) worn on the front left shoulder had two uses; the tapered tip could be used to 'spike' the enemy's cannon by ramming it into the touchhole and snapping it off, and the length of cord could be used to tie up a section of captured horses – the more senior the officer, the longer the cord. The bearskins are made from the non-endangered Canadian brown bear and dyed black.

A macabre weapon, a poleaxe, was carried by the farrier, who rode behind the standard bearer in battle. This was a heavy axe head with a spike on a stout staff. The spike was used to humanely dispatch a wounded horse by severing the spinal column behind the ears and the axe was to cut off a hoof after the horse had died. Only by presenting the hoof with its branded identification number could the officer be issued with a new horse.

This was the result of an army racket revealed at Waterloo. Officers requested a new horse claiming that their horse had been killed but in fact it had been sold on. Ever since then army horses have had identification marks branded on their hoofs.

Household Cavalry horses and soldiers on holiday on Holkham Beach, Norfolk.

The ceremonial saddlery in the foreground laid out for polishing is that used by Irish Guards Colonel of the Regiment, HRH Prince William. Saddles around the walls have the typical deep seat with high cantles of a traditional cavalry saddle with several 'D' rings for attaching kit.

The saddle cloths, ceremonial and campaign, have the identifying regimental badge and motto but that used in warfare is plainer and less conspicuous. In fact, during campaigns soldiers were ordered not to polish anything as buttons, cap badges and stirrup irons could reflect the sun and give away their position to the enemy.

The heavy-duty bridles, with cheek straps of one and a quarter inch width leather, have an integral headcollar which clips together with a stud where browband and headpiece meet. This practical design allowed for removal of the bit when the horses were fed from nosebags during marches.

The double bits consist of a bridoon bit and a state bit (Weymouth) with elegant curved shanks bearing the regimental emblem. The emblem or capstar is also on the horse's forehead and breast plates. For ceremonial occasions when the Queen is present, the commanding officer's charger wears gold braided reins.

The workshop also fits all the cuirasses (body armour) and boots to each soldier, the oak-tanned leather being supplied by J. & F.J. Baker (*see* page 2). Firmins of Birmingham, who have manufactured swords since 1677, supply the cuirasses, elaborate helmets with plumes and the ceremonial bosses which adorn the bridles.

The Queen's Life Guard trumpeter's horse is always a grey so as to be conspicuous for his duties as a messenger during battle engagements. All the other horses are dark bay or black, apart from the drum horses.

The Shire horses which carry the silver kettle drums (presented by Charles II, George III and William IV) are guided by reins attached to the stirrups of the drummer so that his hands are free. The drum horses, named after Greek gods, have the rank of Major and are accorded the appropriate respect by junior officers.

King's Troop Royal Artillery

The King's Troop Royal Horse Artillery is the Queen's ceremonial battery and part of the Household Troops, which was instigated by her father, King George VI in 1947. Its origins are in 'A' Troop of Horse Artillery formed by Royal Warrant in 1793 and then based at Goodwood and Woolwich. The Royal Horse Artillery (RHA) fought with distinction throughout the 19th century campaigns in Iberia, Waterloo, the Crimea, India and Afghanistan and in World War I.

The King's Troop is a mounted unit whose teams of six black horses pull gun carriages for state and military funerals, public performances and salutes which are fired on the occasion of national anniversaries. The thirteen pounder field gun and limber together weigh thirty-two hundredweight and its shells had a range of nearly four miles. Their 'Musical Drive' display performed at the gallop with the massive momentum of horses and gun carriages spectacularly portrays the crushing forces of warfare.

In August 2014 the King's Troop RHA took part in a centenary re-enactment of the Retreat from Mons in which the exact hundred mile route was ridden by a troop of soldiers wearing First World War uniforms, eating the food that would have been issued then and sleeping in canvas tents. A salute was fired from a horse-drawn thirteen pounder gun at a moving service of remembrance on the battlefield at Néry where three of the sixteen Victoria Crosses awarded during the fighting retreat from Mons had been won.

For one month every year the King's Troop also take over the duties of the Queen's Life Guard. Their men and women are also full fighting soldiers and were deployed in Afghanistan.

Their hundred or so horses are stabled in new accommodation at Woolwich Barracks but used to be in St John's Wood where their Top Harness Room housed the splendid gun carriage harness. Around the room are painted the names of the drivers of 'E' Troop Royal Horse Artillery during the Waterloo campaign, the barracks having seen over two hundred years of occupation by horses and soldiers.

Each harness is the responsibility of a small team of soldiers who stand by during routine inspections

'Honi soit qui mal y pense' (Shame on him who thinks evil of it), referring to the Garter Star, one of the oldest symbols of the British Monarchy, is inscribed on the capstar of the bit on this Coldstream Guards bridle. It derives from Norman French and was spoken by Edward III in the fourteenth century on the eve of battle.

The King's Troop Royal Horse Artillery Top Harness Room in the St John's Wood barracks, 1880–2012. The traditional Army Universal Pattern cavalry saddle shows the long felt-covered panels of the tree which distribute the rider's weight over as large an area of the horse's back as possible.

when the back of every buckle is examined and the tiniest fleck of metal polish on leather will mark down the team's record. The best team was awarded custody of a ceremonial sword for the week. Shoe polish is used on the outside of the harness, bridle and saddle to give a high shine and saddle soap or oil on the underside to nourish the leather.

Previous experience with horses is not required for new recruits to be accepted into the Household Division. After twelve weeks of basic army training those selected for mounted duty spend sixteen weeks at Windsor and are taught to ride (the first three weeks without stirrups!) and care for horses. Sword and cavalry drills follow with practice at riding in ceremonial uniform. After the Passing Out Parade the recruit moves to the Hyde Park or Woolwich Barracks to be based there with his horse for two years and instructed by the Household Cavalry's Riding Master. Training continues and soldiers may be deployed to other units depending on the military situation.

The Riding Master is also responsible for buying the horses, known as remounts. Most are sourced in Ireland as youngsters and trained at one of the Army remount centres.

Horses have been our tools of war in battle lending their strength and courage amongst cannon, sabre, rifles and barbed wire in an environment which could not have been more alien to a horse's nature. It is remarkable how trust in its rider and careful training can give a horse the confidence to rise to the challenges of warfare. Training includes teaching horses to lie down for concealment purposes and lying still enough to provide a stable firing platform for a rifle.

An intriguing account of the power of training is said to have been given by Sir Astley Cooper (1768-1841) a brilliant and pioneering surgeon who was also an examiner at the Veterinary College in London. Apparently, on seeing a consignment of wounded horses just arrived home from the Battle of Waterloo, he purchased twelve of the most serious cases. He and his students laboriously removed all the bullets and grapeshot from them and when they had recovered he turned them out into his park. He was surprised to see that every morning they would form an approximate line, charge, retreat and then gallop about.

In 1982 Sefton was one of fifteen Blues and Royals horses on their way to the Changing of the Guard when the IRA detonated a bomb as they passed Hyde Park Corner. Four soldiers and seven horses died and Sefton and his rider, Trooper Michael Pedersen, were badly injured. Sefton suffered a severed carotid artery, shrapnel in the bone and other wounds but recovered to resume his duties with Pedersen. Before his retirement two years later, Sefton often passed the spot where the attack had happened but showed no sign of fear.

Some horses became tolerant to the noise and close engagement of cavalry battle, such as the Byerley Turk (1678–1703) who fought in the Sieges of Vienna and Budapest. He became Major Robert Byerley's charger with the 6th Dragoons at the Battle of the Boyne in 1690 and apparently relished a skirmish. He retired to stud in Yorkshire to become a foundation sire in the development of the English thoroughbred and is buried on Colonel Byerley's Goldsborough Estate.

Warrior, on whom the book and film of *War Horse* is based, was another survivor of active cavalry service with his owner Lord Mottistone (General Jack Seely, grandfather of Brough Scott MBE, racing journalist and jockey).

During his years on the Western Front, Warrior became quite unafraid of shell and rifle fire in spite of many hair-raising escapes where shells killed horses and soldiers all round him. He returned from France after the First World War to enjoy point-to-point racing and hunting in happy retirement. His death, at the age of thirty-three, was announced in *The Times* newspaper and he was buried in the Isle of Wight where he was born. He too, by Lord Mottistone's detailed account, seemed to actively enjoy the adrenalin of warfare.

Copenhagen (1808–1836) the Duke of Wellington's charger was also well documented. A small chestnut stallion whose grandsire was Eclipse, he was a failure as a racehorse but became Wellington's favourite hunter and campaigner. Both of them endured the intensity of the Battle of Waterloo for fifteen hours, the horse apparently unperturbed. Copenhagen spent his retirement enjoying the gardens and estate at Stratfield Saye in Hampshire where he was buried with full military honours at the age of twenty-eight.

Napoleon's favourite grey Arab, Marengo, is surrounded by myth and confusion about his origins and campaigns. It is thought that he was wounded in the Battle of Waterloo, captured and taken to England, probably on the same ship as wounded horses of the Household Brigade who were repatriated to England.

An officer's affection for the heroic mare Sal shines through his memorial to her. Her resting place on a peaceful estate in North Wales was well deserved.

His skeleton is exhibited in the National Army Museum in London.

A horse named Sal is not famous but was undoubtedly an equine heroine. The catastrophe of the Charge of the Light Brigade overshadows the extraordinary bravery of the Heavy Brigade's action at Balaclava in which Sal took part. She was one of 800 British Heavy Brigade cavalry who charged 3000 Russian light horsemen and drove them back. The rout was completed with the help of guns from 'C' troop of the Royal Horse Artillery, forerunner of the King's Troop RHA.

To have also survived the journey of up to six weeks on the sailing transport to and from the Black Sea, let alone the appalling conditions for nearly three years in the Crimea, is miraculous. Her seventeen years of service to the Royal Dragoons (now part of the Queen's Life Guard) was exceptional and her name lives on though her officer is long forgotten.

Horses above the height of fifteen hands high were requisitioned for the Second World War and although no longer used by the British Army as cavalry, they were still needed for transport and to patrol difficult terrain such as in Palestine, Syria and Italy. The Wehrmacht however, employed more than three million horses between 1939 and 1945 as Hitler's offensives had run ahead of mechanisation production and the supply of petroleum. The Red Army is thought to have had at least that number as well.

In the First World War, 130,000 Australian horses (Walers) were requisitioned: many were from outback stations and came with their owners who signed up to fight. The surviving cavalry from Gallipoli were sent to Egypt and crossed the Sinai desert to successfully take Beersheba and Damascus. At the end of the war and after years on harsh campaigns together, the soldiers were told that their horses were not to be repatriated. None of the 130,000 came home.

The pity and hopelessness of the horses caught up in conflict is portrayed in one of the most famously moving war paintings by Fortunato Matania (1881-1963). It was commissioned as a poster in the First World War for the Blue Cross animal welfare charity and shows the extraordinary bond between soldiers and their horses. *Goodbye Old Man* depicts a soldier bidding his stricken horse farewell surrounded by scenes of devastation. Henry Chappell's (1874-1937) poem 'The Soldier's Kiss' complements Matania's painting and is equally poignant.

The Household Cavalry Foundation is a charity which supports the soldiers and assists in finding placements for retired horses. The Horse Trust in Buckinghamshire, founded in 1886, claims to be the oldest horse charity in the world and provided the first motorised ambulances to transport wounded horses from the frontline during the First World War. In 2017, the Trust welcomed Viscount, a 17hh Irish gelding, on his retirement after serving a record twenty-one years and eleven months with the Household Cavalry Regiment.

During both world wars many charities worked with the Army Veterinary Corps to improve the conditions for millions of mules and horses. It was the pitiful sight of cavalry horses abandoned after the war to hard labour which inspired Dorothy Brooke to establish the Brooke Hospital in Cairo in 1934. Its assistance to working equids abroad continues and has inspired charities such as World Horse Welfare, the Society for the Protection of Animals Abroad (SPANA) and many others to support vital welfare for the estimated hundred million mules, donkeys and horses still working today.

The Animals in War memorial, Park Lane, London, designed by David Blackhouse in 2004 commemorates all the animals who died in 20th century wars, including the eight million horses estimated to have died in the First World War alone.

CHAPTER THIRTEEN

Coaching and Carriage Driving

In the Golden Era (1815-1840) of coaching the soundscape of our environment must have been entirely different from today: hoofs striking sparks on cobbled yards, stable lads shouting, horses whinnying and carriages rattling through the towns. The post horns would sound as the mail coach approached at speed for a change of horses and in the villages the chink of churns in the milk float, the jingling of harness and the grind of agricultural wagon wheels under heavy loads were as common as the cock crowing.

The only way to travel from place to place up to the beginning of the 18th century was by foot, on horseback, in a horse-drawn vehicle or on a coastal sailing ship.

Although transport by canal and railway became more widespread in the 19th century, there were enormous numbers of horses in and about town, one for every ten people approximately. As late as 1900 there were 46,000 carriage horses in London, according to the equestrian artist Lionel Edwards who lived in London at that time, and horse cabs were licensed in London until 1943.

A vast range of horse-drawn vehicles was needed for getting about in town and in the countryside: such as hackney cabs (first licensed in 1610), gigs, barouches, Broughams, phaetons, landaus and Hansom cabs, but for long-distance travel, a coach drawn by four horses was really the only choice. For private owners and coaching companies the harness not only had to be functional and safe but also immaculate. Coaching etiquette for the presentation of horses, harness, carriage and the coachmen's livery was minutely observed. Similar attention to detail applies to the judging of turnouts in the show ring today. Owner drivers wear grey top hats, otherwise black toppers are correct which is why the Duke of Edinburgh is often seen in black as the Fell ponies he drives belong to the Queen.

The coach is thought to have been introduced to the court of Queen Elizabeth I from the Continent in the 16th century and in 1677 the Worshipful Company of Coachmakers and Coach Harness Makers was established in the City of London (*see* Livery Guilds page 11).

British coachbuilders developed the design further in the late 18th century with the Collinge axle which secured the wheel and lubricated the bearings with oil stored in the hub cap. The invention of elliptical steel leaf springs (patented in 1805 by Obadiah Elliot) was an improvement on the 'Cee' springs with leather braces which caused passengers to feel seasick from the motion. The new springs made the ride considerably more comfortable and with the improvement in roadbuilding by Telford and Macadam, the coach bodies became lighter and faster.

Mr Bridges with his groom on the road to London by Benjamin Herring Snr, 1828, showing two horses in tandem harness drawing a 'cocking cart' – the sports car of its day.

Stage coaches, which first appeared in the 1600s, were pulled by four horses with room for four persons inside and at least ten on the roof. Passengers were said to write their wills and offer prayers before embarking due to the hazards from bad roads, highwaymen and accidents.

Royal Mail coaches replaced the post boy riders in 1782. These coaches ran to the minute and carried a locked timepiece which was opened to record the time at each change of horses, giving rise to today's saying 'to pass the time of day'. It carried up to seven passengers and a guard armed with pistols.

The post horn, a circular horn without valves, had nine different calls, most of them for warning traffic ahead to 'Clear the road!' as the Mail coach had priority on roads and at tolls. It is said that the coach ran over the legs of a man who had failed to get out of the way in time, severely disabling him. He took the Royal Mail to court to obtain damages but was unsuccessful and, to add insult to injury, the court prosecuted him for impeding the mail service.

Famous stage coaches, like railway engines, had names and reputations. A 'stage' was the distance between inns and, depending on terrain and the road conditions, ranged from five to fifteen miles. In 1834 'The Wonder' travelled the London to Shrewsbury route of 150 miles daily in fifteen hours, including three stops for meals, for a fare of £1.

The coach carried a guard and sixteen passengers and weighed one and a half tons. This particular coach was painted by Charles Cooper-Henderson (1803-1877) whose work depicted coaching harness with great detail and accuracy.

Writing about the London to Brighton coach run, Walter Godden, author of Ye Coach Horn Tootlers, in 1893 reported that after going down several hills he could fry a couple of rashers of bacon on the drag shoe (an iron slipper into which the wheel rim fits to stop it turning and used as a brake on steep downhill slopes). One of the last coaches to run between London, Worthing and Brighton was 'The Venture' owned by the flamboyant American, Alfred Gwynne Vanderbilt who often drove it himself, dressed immaculately and wearing a grey top hat. His stone memorial by the side of the A24 at Holmwood, raised by his friends after his death in the Lusitania, torpedoed in 1915, is a lonely reminder of the coaching age.

Stage coach horses had a short working life often lasting only four years. The origin of 'nightmare' was the practice of using lame horses in the coach team at night when their condition was less obvious. 'Dropping off to sleep' referred to tired drivers who would nod off and then fall from their high seats. Keeping warm in such an exposed position for hours at a time was difficult and coachmen's boots were large and made of boiled leather.

Coach horses could be both driven and ridden. A postilion would ride the nearside horse and lead its offside pair. The ridden horse would wear a saddle

The Arlington Court harness room (see page 83). The Earl of Onslow's state harness is on the right of the saddle horse and behind it are several whip reels on the wall for storing driving whips. Original coaching accessories are kept here too: copper hot water containers for warming the passengers' feet, leather document cases, top hat boxes and a collection of coach horns.

Rachel Lancaster (whip) and Samantha Jones (backstepper) finishing the dressage phase at the Chester horse driving trial with hackney pony Heartland Power Stroke (see Competitive driving, page 84). The harness is a Zilco classic set made of biothane with an Empathy collar, curved to allow the shoulder to move more freely. The bit is a jointed butterfly Pelham bit which is gentle on the mouth.

Prince Philip at Windsor driving friends on his 70th wedding anniversary, 20 November 2017.

instead of a pad, the traces passing under the rider's lower leg. An iron leg guard was strapped to the outside of the postilion's right leg to protect it from the pole. It may be that this iron guard contributed to the postilions' occupational hazard of popliteal aneurysms, a fatal complication of the artery behind the knee, until John Hunter (1728-93) devised a curative operation.

Coaches and carriages were not only for transport but among private owners were a sign of their social rank. Many were elaborately ornate with painted scenic panels, silver and gilt fittings, coats of arms and the livery of the owner or company. Their horses and harness were equally on show. When Florence Nightingale was a child she toured Europe with her parents and sister in a huge coach designed by her father. Besides the family of four, there was also room for two maids, a footman and a messenger. It was drawn by six horses led by three postilions and after many months returned to England transported across the Channel in a sailing boat.

Original carriages and harness may be seen at the Royal Mews, the National Trust Carriage Collection and in stately homes and museums. A national register of all coaches and carriages in Britain is being compiled by the Carriage Foundation.

National Trust Carriage Collection at Arlington Court, Barnstable, Devon

The colonnaded stable block houses the carriage museum and a fine harness room which is its central focus. It is a cool dark room opening onto a passage leading to looseboxes on one side and the carriage house on the other. A cast-iron range was lit in winter and provided the only heat.

Two sets of 19th century state harness with the silver coat of arms of the Earl of Onslow hang alongside the much plainer modern harness used for wagonette rides for the public on the estate. The modern harness is of the traditional urban driving design with Liverpool bits, some with an 'elbow' shank to prevent the horse from chewing the rein, and with a thin metal bar connecting the lower ends of each shank to stop them catching on the adjacent horse's harness.

The English collars for carriage work are leather lined and the harness includes the 'bearing rein' abhorred by Anna Sewell in *Black Beauty* (1877) which was so effective in the RSPCA's early campaign. This is an adjustable rein which runs up from both sides of the bradoon bit, through a ring fixed to small straps (bradoon hangers) on the headpiece and back to a hook on the saddle or fly pad. Its primary function is to keep the horse's head

THE TACK ROOM

An obstacle on the marathon course negotiated at a fast trot. This shows the skill of the driver on the reins; he has already turned the leaders while the wheelers behind them are on a different track to bring the carriage safely through the gate before they also turn.

up and under the control of the driver for safety; a horse cannot bolt effectively unless its head is down and its jaw set against the bit. It is also argued that it can prevent a horse falling to its knees if it stumbles. However, if the rein is tightened too much to present a more dashing profile, the head is raised unnaturally and causes considerable pain.

To the carriage driver, the whip is the equivalent of the rider's legs on a ridden horse and is an essential part of all driving harness, used to guide and encourage the horses. Different types of driving, for example trade turnouts or private driving, require different whips.

A competition team whip (for four horses) has a stick about five feet long and a thong and lash of about twelve feet. Many of these bow-top whips were finely crafted, decorated with silver butt caps and ferrules and their leather or silk covered handles matched the inside furnishings of the carriage. Traditionally, the stick would have been made of seasoned holly, chosen for its natural knobbles which prevented the thong from slipping down when correctly folded. The plaited leather thong was made on a tapered core of baleen (whale cartilage) which shaped the thong into a bow or arch. The thong was attached to the stick with strips of calfskin wound through the plaiting, supported by split goose quills and bound tightly to the holly with black linen thread. To strengthen the joint between the baleen core and the tip of the stick, split goose quills six or seven inches long were placed over the calfskin strips and bound again with black thread. The other end of the thong was finished with a lash of leather or Chinese silk cord. Sometimes the last fitting left in a stripped out harness room is an empty wooden whip reel pinned high up on the wall. The thong was positioned in the groove with the stick hanging vertically to keep the arch in the right conformation.

Modern driving whips made of synthetic materials and with telescopic handles are preferred by some, such as Boyd Exell who drives his team with two different whips: one for the wheelers and another for the leaders.

This National Trust collection has over fifty coaches and carriages, including the 18th century Speaker's State Coach from the House of Commons. These are the thoroughbreds of the carriage world: gilded and varnished, full of the elegance and grandeur of the Golden Age with impeccable provenance, their interiors upholstered in brocades, silks and Morocco leather with embroidered monograms. Some may once have had the dust of the Grand Tour on their wheels but are now dormant and embalmed by conservators. Many were exported and lost. In non-temperate climates without the correct relative humidity (65%), carriages quickly deteriorate due to shrinkage of the wood felloes (curved rims) on the wheels so that they will literally fall apart.

Fortunately, the Arlington stable yard escaped conversion into a tea room and shops, unlike those at Berrington Hall, Castle Howard and so many others, because it was a short distance away from the house – too far for the visitors to walk – and the working stables and tack room have survived to show how it was in the 19th century.

Competitive driving

Apart from driving for pleasure there is plenty of competition both in the show ring and in driving trials. Classes include singles, pairs, tandems and teams for a wide range of horses, ponies and vehicles – from governess carts pulled by Shetlands to traditional trade turnouts with heavy horses. Each turnout has a harness to match the type of vehicle and the number and arrangement of the horses are too diverse to be individually described here. Events are held by the many driving clubs throughout the country and entered by children, grandparents, novices and professionals from all walks of life.

Combined driving trials are modelled on the ridden Three-Day Event (*see* page 108) and were founded by HRH Prince Philip in 1964 when he was president of the International Equestrian Federation (FEI). Sir Michael Ansell's committee drew up the rules and the first international competition was held in Lucerne in 1970. Prince Philip also commissioned one of the first purpose-built cross-country carriages (from Bennington, holders of the Royal Warrant) with a steel chassis because the traditional old carriages were not robust enough for trial conditions.

A combined driving competition has three phases including presentation: dressage, a cross-country marathon with obstacles and manoeuvring in and out of cones in an arena. The dressage test consists of a set sequence and is judged on the appearance and correctness of the turnout, and the obedience, paces and harmony of the horses. The marathon phase is timed over different distances (five to twenty-two kilometres), depending on the standard of entrants, with obstacles, set pace sections and mandatory halts for vet checks. The third slalom-like phase in an arena is a timed drive to test for suppleness and accuracy among a pattern of cones, each topped by a tennis ball which must not be dislodged. A crew may consist of the driver, a groom or 'backstepper' to help balance the vehicle round corners and a navigator (for the marathon course). Carriages for the cone and marathon phases are purpose-built cross-country vehicles with a steel chassis and wheels, robust springs and disc brakes and have little in common with traditional designs. Carriages for the dressage phase may be either modern or traditional such as a phaeton.

The harness is of standard design appropriate for the vehicle and number of horses used, except that

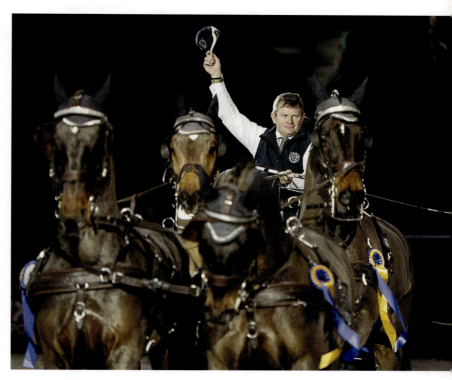

Boyd Exell (see page 86) winning the FEI World Carriage Driving competition in Sweden in 2014.

for driving trials leather is usually substituted with synthetic biothane (Zilco, Australia), or beta which looks like leather but is lighter, easier to maintain and stronger. Bridles with winkers and a variety of bits are used.

The BSE outbreak and the 1996–2005 ban on cattle over thirty months old from entering the food chain reduced the availability of large mature hides for harness-makers. Older cattle are now all tested and if negative do now enter the food chain. Hides from Germany which are long enough for traces are imported to cover any shortfall.

FEI World Cup competitions are popular spectator events which combine some of the marathon and cone elements. Driven at top speed it is a thrilling spectacle, and the Extreme Driving at Olympia is a World Cup qualifier. It brings spectators to the edge of their seats to see the astonishing agility of a team of four-in-hand manoeuvre a figure of eight through narrow gates and cones between full gallop sections. It looks as impossible as threading a camel through the eye of a needle, yet miraculously they emerge at speed with the obstacles intact.

Boyd Exell's large and comfortable tack room which doubles as an office. Harness with breast collars for each team of four is easily transported from the tack room to the horses on wheeled frames. Lisa Mitchell (pictured here) has been Boyd's long-term navigator on the marathon phase.

Boyd Exell

Boyd Exell has an extraordinary talent for competitive horse driving trials and holds a record number of four World Championships, nine British National championships, seven FEI World Cup titles and numerous other awards.

Born in New South Wales, Australia, he was keen on horses from a young age, though as his parents had little interest in them he became an apprentice engineer. He was introduced to a local carriage driver and by the age of sixteen he had won his first national championship.

He established a base in the UK for more than twenty years before moving from Leicestershire to Holland to be closer to the continental circuit where driving is more competitive. For his teams, in which horses are matched for colour, markings and size, his preferred choice is a Dutch warmblood crossed with an English hackney, or a Lippizaner broken in at three years old, and gradually brought up to international competition standard over the next two or three years.

Boyd's tack room is a large and comfortable room inside a complex of barns and stables. The amount of harness needed for a team of four is considerable and would normally need a lot of fetching and carrying between tack room and horses. To speed up the harnessing process, each horse's harness is stored on a wheeled metal frame which can be pushed out to the yard alongside the team. Boyd was inspired with this idea while in a restaurant one day and saw the plates and dishes stacked and easily moved from kitchen to dining room in a similar frame.

Boyd's preparation of each team of four (and a fifth reserve horse for FEI competitions) is meticulous. He rides several of the thirty or so horses a day in quiet concentration, schooling each through dressage techniques to increase its balance and suppleness and finding the type of bit which suits each individual. Just as important is his feel for the horse's temperament: whether it is suited to being a wheeler and comfortable with the carriage pole alongside or has the dash and courage to be a leader. Horses learn to work in pairs before being 'put to' as a team of four. Boyd will also work several of his teams every day as it takes weeks of practice for a team to come together as one happy family. To keep them from getting stale the horses also hack out and sometimes go hunting.

Carriage driving requires great sensitivity and dexterity, the reins being two-way channels of communication for a constantly nuanced dialogue between the driver and each of his four horses. Handling the reins of a team has been likened to playing a harp, and applying the whip accurately in the right way to the right horse also takes much practice. One way to improve this skill at home is to simulate the team by setting out four candles on the floor of a long corridor the same distance away from a fixed point (where the driver would sit) as the four driving saddles or pads would be when hitched to the carriage. The object is to lightly knock down each candle individually with the lash of the whip without upsetting the others.

Great skill in anticipation and timing is needed, especially at speed, because of the total length of a four-horse team and carriage – the pole alone measures ten feet.

Boyd is a modest man, crediting his success to the equine and human teams. He is also Britain's four-in-hand coach and continues to train many international competitors. It is his hope that one day driving will be an Olympic discipline. 'I try to convey to novices that successful driving is about feel and feedback for each individual horse in the team, followed by steady practice', Boyd says between receiving constant calls on his mobile, the ringtone for which is The Post Horn Gallop (also played at Leicester City home matches, though I never discovered whether he was a football fan as well).

George Bowman

George is a pioneer of driving, having been in the sport for forty five years and has been British four-in-hand champion nineteen times. He has been a member of seventeen World Championship teams and is still competing in international competitions. 'It's like chasing a rainbow', George is quoted as saying, meaning that with so many variables, it's difficult to achieve the perfect performance. 'And adding in a new horse can set back progress for some time until the team works together again.'

George began by rodeo riding in Canada but after returning home he had a serious accident at work and temporarily lost the use of his legs. While he was convalescing he took up driving and never looked back. In 1972 George drove a coach from Edinburgh to London, no mean feat considering the traffic as well as the logistics for four horses, and in 2009 he and family members drove a stage coach across Ireland, raising money for children's charities.

The carriage driving community has a strong reputation for friendliness and inclusivity and in competitions mutual support is part of the convivial atmosphere. Many events are ideal for families to attend, often with camping on site, and held on estates in beautiful parkland. It is an absorbing sport for any type of horse, for drivers and grooms who are young, old or have special needs and which can be enjoyed outside or in an arena all the year round. It is said that carriage drivers never retire; two of the finest examples are ninety-six year old HRH Prince Philip and eighty-three years old George Bowman, both of whom are still holding the reins.

George Bowman, Four-in-Hand National Champion nineteen times, in his famous yellow livery on the marathon course in Cheshire.

CHAPTER FOURTEEN

The Circus

The razzle-dazzle of trumpets, costumes and coloured lights, the Big Top lit up like an enchanted palace, the excitement, magic, laughter, the whole experience intensely brilliant; this is the circus and as seductive as the Pied Piper. It is where being crazy, funny and daring is allowed, a never-never land, a parallel universe of exotic colour and imagination, where the laws of the universe are suspended.

But the circus is far more than a children's entertainment with its long tradition of skills and a way of life often lived by many generations of the same family. It is a complex, hierarchical and close-knit community with its own lore and internal discipline. The clowning and death-defying acts offer an almost therapeutic experience for the audience with comedy and pathos that was recognised by the ancient Greeks.

The juggler, derived from the French 'jongleur', meaning a joker, harks back to the court jester and reflects the wide variety of skills each circus performer can display. Yasmine Smart, grand-daughter of Billy Smart, is not only an internationally renowned expert in training horses but also a high wire walker, a trained costume couturier and a champion arm-wrestler. Her mother, Olga Smart, was a trapeze artist and her grandfather crossed the Thames on a wire in 1951. There was always a wire outside their tent for practice on which Yasmine began when she was ten years old, though she had switched to working with horses by the time she was twelve.

Circus performers dare to be different from those of us who go to school, live in the same place for years and try to find a job, a mortgage and a home.

The jossers (those who work in the circus but were not born within its community) have a foot in each camp: a choice between the ordinary world and the extraordinary circus. They oscillate between being 'them' the outsiders and 'us' the true children of the circus with sawdust in their shoes. For those born into it, it is a compelling way of life which offers a freedom of sorts – the freedom of expression in perfecting a talent and a unique environment in which artists can dedicate themselves to the art of performance. But the rigorous discipline of constant practice is a hard taskmaster. There is little job security because of seasonal contracts and it is a livelihood which does not tolerate old age easily.

There is a purity in circus entertainment which is dependent solely on human ingenuity and physical skill with no added trickery from computer-generated imagery for special effects or stunts. This is reality, not the artifice of games on an electronic screen but created through hours of practice.

Nor in a modern circus are there any short cuts in the training of any animals which are part of the show. Success is based on a system of reward and the willing cooperation of the animals to perform. Anthony Hippisley-Coxe (1913–1988), a great authority on the circus, was sceptical about the claim that animals could be taught to perform without force. To test this he bought six domestic cats and trained them without cruelty to carry out a number of sophisticated tricks to music. Ever after

Spotted delight with Appaloosa ponies and Dalmatians at the Giffords Circus 'Thunder' show.

they would spontaneously perform if they heard their familiar theme music, purring loudly throughout and convincing himself and an RSPCA inspector that animals can enjoy such work.

Medieval seasonal fairs, menageries and freak shows entertained with jugglers, acrobats and performing animals. By the 18th century fairs were causing frequent public disorder and, combined with rivalry from the theatres, this led to restricted performance licences. Southwark Fair in London, illustrated by a vivid series of engravings by William Hogarth (1697-1764), was charged with immorality and riot and closed in 1763. In London and also Paris the fairgrounds went into decline and the performers migrated to fledgling circuses.

The modern circus is credited to Philip Astley (1742-1814), a British cavalry officer who had fought with distinction in the Seven Years War. On leaving the army he was presented with a white charger, named Gibraltar, with whom he started his new career as a riding master and entertainer. Taught by the élite riding masters of their day, Cornet Floyd and Domenico Angelo, Astley developed his cavalry skills to become a brilliant trick rider, performing stunts so extreme that they are rarely seen today: supporting himself only by his arms between two horses galloping at full speed or riding on his head firing a pistol and lying across three horses at full speed, to name just a few.

He married an equestrienne and at their premises in Lambeth, London, in the 1760s he would give riding lessons in the morning and entertain the public in the afternoon. Later, clowns and jugglers were introduced in between the equestrian acts.

Astley realised that a circular arena surrounded by tiered seating was the optimum arrangement for a public show and that a diameter of thirteen metres best suited a horse at a collected canter. Furthermore the centrifugal force made it easier for vaulting and for trick riders to balance. Before Astley, mounted performances were held either in a rectangular manège or in a linear field where this advantage was undiscovered. The diameter of thirteen metres (or 42 feet) has been the international circus ring standard ever since.

On one occasion when passing by a parade in London, he was able to assist King George III in calming down his horse which had been frightened by cheering crowds. Astley was granted a Royal Command performance followed by a fourteen year contract to train the cavalry horses to tolerate the noise of trumpets and explosions. His move to near Westminster Bridge and the development of Astley's Amphitheatre in 1780 for his shows, enhanced his fame and led to international invitations. Astley erected many provincial amphitheatres and at least eighteen circus buildings in Europe. His Amphitheatre Anglais (later the home of Franconi's Circus) in Paris was typically splendid with dozens of huge chandeliers lit with jets of flame. Not surprisingly, as most were built of wood, many of his amphitheatres were burned down several times and rebuilt.

After a flamboyant life, and leaving his indelible mark on circus history, Astley died in Paris in 1814. In 2015, a statue of him (by the sculptor Andrew Edwards) was erected in his birthplace of Newcastle-under-Lyme, Staffordshire. From these origins in London, circuses, which became tented after 1840, are now found across the world. In 2018 it will be the 250th anniversary of Philip Astley's first circus in London.

Horses have been associated with modern circuses from the beginning and are well documented in a large body of art inspired by the circus. Toulouse Lautrec, Dame Laura Knight, Edward Seago and Picasso are some of dozens of artists who have captured their ephemeral brilliance.

Exotic wild animals have thrilled audiences in Britain with spectacular entertainment for at least a hundred years. Since the Second World War, several animal rights organisations, such as PETA (People for the Ethical Treatment of Animals) in the USA have campaigned for the abolition of wild species in the circus, even though they are born in captivity. Such pressure persuaded many councils in England and Wales to refuse licences to travelling circuses with performing wild animals and many disposed of them in order to survive. In Britain the 'Wild Animals in Circuses' bill banned their use from December 2015.

Gifford's Circus

Gifford's Circus has horses at its heart, along with top artistes from all over the world. Nell Gifford realised a childhood dream to have her own circus when she and her husband, Toti, opened their touring Victorian-style show at the Hay-on-Wye Festival in 2000. Since then, Gifford's has become an essential part of Middle England's summer and won several awards.

Nell swims against the tide of convention with ease; she graduated from Oxford with a degree in English and, with barely any experience save a month with Circus Flora in the USA, went to work as a groom in circuses in Britain and Europe, learning all the way. Mucking out horses winter and summer, selling tickets and ice creams and living in threadbare, leaking caravans were hard beginnings. Later she joined the French circus Santus which was touring in England and caught a glimpse of the professionalism and celebrated status of European circuses. Nothing is easy for a josser though and the years of back-breaking work with little money or encouragement would have put off all but the most determined. When Nell was nearly eighteen, her mother, who also loved horses, had a fall in which she was critically injured. She recovered but was severely incapacitated for many years until her death. Despite this, Nell's exceptional perseverance and imagination held firm throughout. She and Toti began with little more than a vision of a village-green style of circus for families, working on a shoestring out of a garden shed as an office and supported by friends. As it developed they travelled in Europe and as far as Moscow to engage the best

Nell Gifford with Tsijimke (Jim) her Friesian stallion.

Gifford's Circus. Yasmine Smart (left) and Nell Gifford (right) introducing the Welsh ponies.

acts they could find, returning to Gloucestershire to plan, rehearse and hand-make scenery and costumes over the winter.

Toti's farming background and talent for making just about anything the circus needed made sure of Gifford's reputation for authenticity and smartness. Original vintage circus wagons were rescued from ditches and dilapidation, restored and painted and winter quarters built. He also designed and built a unique travelling tack room. It is box-shaped with a tough frame which can be hydraulically loaded onto a lorry. The canvas canopy of the horse tent rolls up and is stored inside it and the horse stall sections are transported on a flat trailer.

The fixtures on which the saddles, bridles and vaulting rollers hang in the tack room are designed to withstand an angle of forty five degrees as it is winched onto the back of a lorry. Even the feedbin is secured with brackets and a lockable lid so that nothing can spill. The 'pull-down' procedures of decamping from a site are so well rehearsed that the horses, their stabling, fodder and tack room were all packed up and on the road in twenty-five minutes.

Most performers are on short-term contracts for one season and then move on to other circuses. I had corresponded with Tamerlan Berezov from Ossetia on the Russian border who joined Gifford's for two seasons. He was a seasoned Cossack-style rider and horse trainer who travelled the world with Elena his young wife, two little children and his handmade, rawhide Cossack trick saddle. After the performance was over – in which he had demonstrated his horsemanship and she had presented her flock of obedient doves – they entertained us outside their caravan. She rolled out biscuit pastry on a little table in the open air and baked them while we chatted in broken English and in French which is the common language of circuses. Brian the goose kept a close eye on us nearby.

The circus show vibrates with energy, colour and loud music. Clever ponies demonstrate their talent jumping, pirouetting and bowing, and athletic bareback riders vault and balance in pyramid formation.

In contrast to these impressive feats there was also a contemplative one when the rhythm changed: a simple but memorable performance. Toti walked behind one of his huge Shire horses guiding it on long reins and Brian (the goose) processed around the ring in single file accompanied by a folk tune on the violin. The horse and Toti were harnessed and dressed for ploughing and Brian, the symbol of a forgotten rural domesticity, followed along behind. They were a striking pageant, in a time capsule, poignant and beautiful. Nell also takes part in performances, usually riding side saddle and their seven-year-old twins are quite at home in the ring.

In 2009 after the show I went to meet the great Yasmine Smart who was spending a season with Gifford's, she and Nell having worked together in 1998 at Circus Roncalli in Germany.

We met in the Gifford's Circus tack room surrounded by ostrich feathers stored in tubes, decorated saddle cloths and pads, diamanté bridles and vaulting rollers with strong leather handles. Two sets of white harness were for the matching pair of pretty Section A Welsh ponies. Special saddles, such as Nell's side saddle and Tamerlan's Russian trick riding saddle with a pommel handle, extra straps and three girths, were kept in their caravans.

In Yasmine's honour, the theme that year was 'Yasmine, a Musical' depicting the stages of her life in the circus. Jugglers, trapeze artists and acrobats played out the scenes to the live accompaniment of music composed for the show, the score of which was included in the audience's programmes – a typical Gifford's touch of detail and imagination. The show ended with Yasmine demonstrating her liberty act with three young Arab horses trained by her at Folly Farm in Gloucestershire over the winter.

At first, without her ringmaster's glittering costume and make-up, I didn't recognise her. Her quiet manner and inner intensity were powerful. She seemed to focus a magnetism which perhaps is her way of communicating with her troupe of stallions in the continental circuses where she works. We talked about the classical equitation techniques adapted for the circus by Henrik Jan Lijsen who wrote the definitive Dutch training manuals, about preparing the Gifford's ponies for performance, how new ones are bought in every autumn to keep each year's show fresh and how much they can learn before the touring season begins in mid-May.

The Gifford's Circus ring of twenty-seven foot in diameter, smaller than the standard forty-two feet, can comfortably accommodate three ponies at once, enough to demonstrate the four traditional equestrian disciplines of the circus:

High School, which includes elements of classical dressage as well as the Spanish Walk (excluded in FEI dressage tests), in which the foreleg is extended horizontally, and is ridden to music. (Dressage to Music had its origins in the circus).

Jockey, or Rosinback, is a performance where the horse, wearing a bridle with fixed reins attached to a girth, canters anticlockwise round the ring while the performer stands on its back.

Voltige is the art of vaulting onto the horse. It can be combined with rosinback work on a horse cantering steadily as before.

Liberty is a performance in which the horses, which may or may not have bridles, have no physical attachment to the ringmaster. Yasmine Smart is an expert in liberty work (and haute école) using only words and the movement of a whip from the centre of the ring to indicate her instructions. Complete concentration and communication between the horses and their trainer is required for this, particularly as they are so closely surrounded by the distractions of the audience.

The popularity of circuses in Britain has dipped due to TV and video games, health and safety legislation, the animal rights lobby and the diminishing cohesion of the extended family for which entertainment has fragmented into being generation specific. Chipperfield's Circus, Bertram Mills and Billy Smart's Circus, each of them giants in the circus world and once commanding audiences of millions in their big tops and on television, have all closed.

Tack room, portable horse stables and hay neatly stowed for the road.

Perhaps as a backlash to processed electronic entertainment, an appreciation of creativity and traditional skills is re-emerging. Circus schools are popular and teach physical skills such as juggling and acrobatics, often interwoven with performance arts such as poetry reading and theatre, but without any performing animals. This new movement is for groups or individuals and has an inclusivity which lends itself to educating and developing new avenues for young people but so far, these are unlikely to grow into circuses.

Gifford's Circus, although a modern organisation, keeps the best of tradition alive. The horses at its centre are trained with respect and kindness. They are part of this special family. Force is never used and no horse would consent to perform in a circus unless it was content to do so. Gifford's Circus provides yet another canvas in which equine versatility and beauty can be appreciated. It has been home to horses who might otherwise have fallen on hard times, as well as cleverly promoting rare breeds such as the Eriskay ponies, an ancient Hebridean breed classified as critically endangered by the Rare Breed Survival Trust. In the autumn the ponies are either kept or go on to make good quiet horses for equine therapy and the Riding for the Disabled Association (*see* page 193) or go to private homes. Those that stay are turned out to grass to enjoy the winter together at the circus farm.

Circuses without animals generally have matting on the ring floor. Those with animals have sawdust and a characteristic smell which adds to the sense of intimacy and participation experienced by the audience. The vast scale of spectacles without animals such as Le Cirque du Soleil and the Chinese Circus provide an entirely different and more remote style of entertainment.

Gifford's traditional village circus invites anybody and everybody to dip a toe into the almost overlooked concept of community, creativity and fun. It also provides a delicious, home-cooked, after-show dinner with local ingredients in its tented restaurant in which the public can meet the performers and share its international family atmosphere.

Nell riding side saddle in 'Any Port in a Storm', 2017.

This infusion of energy, quality and independence to an old tradition revives the faded reputation of the circus and is Gifford's gift to the British public.

CHAPTER FIFTEEN

Classical and Modern Dressage

Classical dressage

Dressage in Britain probably reflects the last four hundred and fifty years of our attitudes to the horse more clearly than any other equestrian discipline.

Classical dressage, which was established in Europe, principally in Naples, in the early 16th century, is the foundation of modern dressage. Also known as haute école (from the 1850s) it includes training the horse to perform advanced manoeuvres known as the 'airs above the ground' such as the capriole and courbette. These are controlled leaps during which all four hoofs leave the ground and, in the capriole, both hind legs also kick out simultaneously. Although central to the tradition of classical dressage, airs above the ground are not required in modern Grand Prix dressage competitions.

A brief collected canter through history describes the development of classical dressage and the influence of William Cavendish, Duke of Newcastle on modern dressage.

Classical dressage, already established on the Continent, was embraced by the English social élite of royalty and peers of the realm in the 16th century. It was valued for its application to training horses for the cavalry but became prestigious in its own right as a high art of equitation which reflected the skill of its rider. While the schooling exercises for suppleness, agility and obedience improved the performance of the horse on the battlefield, the 'airs above the ground' with their spectacular controlled leaps, were never of practical use in the mêlée of war. In any case, the time and expense needed to train a number of horses to this advanced degree would have made it an unfeasible strategy.

Henry VIII (1509–47) encouraged dressage riding as part of court life and, as it became more popular in Elizabethan times, it was expected that high-ranking men should be proficient in dressage riding as well as in fencing, dancing and other cultural pursuits.

In 1516, the writings of Xenophon, a Greek cavalry officer in 5BC, were finally published (in Latin and Greek). His manual, The Art of Horsemanship, details the first descriptions of the bits, bridles and training methods for horses, particularly for the supping exercises, balance and collection of the horse which are familiar to us today. He advocated a sympathetic approach to training based on reward rather than punishment, a method which in spite of its publication was to be ignored by Italian dressage masters in the 1500s.

In the 16th century a form of classical dressage was taught in Italy at the Academy of Naples by Count Cesare Fiaschi. He and his most famous pupil Federico Grisone, a Neapolitan nobleman, both believed in training by force with barbaric methods, such as the use of spiked nosebands. Grisone's book, Gli Ordini di Cavalcare (The Rules of Horsemanship) published in

A tapestry based on Abraham van Diepenbeeck's illustrations in Cavendish's 1658 'A General System of Horsemanship'. The horse, tied to a pillar, is being guided into the correct position for the 'levade' by Cavendish using two wooden switches with metal points called poinsards.

1550 was translated into English on the instructions of Elizabeth I who was keenly interested in horses. Exhibitions of classical dressage manoeuvres were a popular entertainment between bouts of jousting at English tournaments and it was sufficiently familiar in Elizabethan life for Shakespeare to refer to the art of dressage in his plays.

The popularity of classical dressage in Europe led to the establishment of a stud to develop a lighter type of horse using Spanish breeds at Lipizza (now in Slovenia) in 1580. From here the Lipizzaner breed would eventually thrill the equestrian world with its beauty, dancing to Handel and Strauss under the brilliant chandeliers of the Spanish Riding School in Vienna.

Fortunately, Grisone's successor, Pignatelli, and his pupils, Monsieur St. Antoine and Antoine de Pluvinel, were not convinced that training horses by force was best. De Pluvinel returned to France from Naples and founded the first French riding academy in Paris in 1594 to teach his own enlightened methods in equitation and training. He believed in the importance of fitness, lunging and gentle handling to train horses to perform the pirouette, passage and piaffe as well as the 'airs above the ground'. Monsieur St. Antoine came to England to instruct Prince Henry in the early 1600s. Henry was a keen participant in classical dressage as well as in the breeding of horses and substantial gifts of mares and stallions, chiefly from the courts of France, Spain and Italy, filled the stables at the Royal Mews.

William Cavendish (1592–1676), later Duke of Newcastle, had also been a pupil at the Academy of Naples. He returned to England and built a magnificent riding house at Welbeck Abbey in 1620. Cavendish was passionate about training horses, not just for the cavalry (he became Commander in Chief of the Northern Army for Charles I) but for the development of the horse and rider for equitation. Cavendish became a privy counsellor to Charles I in 1639 and Riding Master to the young Charles II. He was a dazzling cavalier: popular, wealthy and cultured, a poet, playwright and musician. Although his training methods were reasonably forceful, he realised that horses had good memories and that therefore ill-treatment could be disadvantageous. He endeavoured to understand the way in which horses learnt and to show that classical dressage was an art. He also considered that the principles and practice of such training might serve as guidelines for human behaviour and for living a noble life.

Illustrations show clearly the tack he designed and used, such as the running reins from the girth through rings on the cavesson noseband to the hand. The saddle had a wooden tree with a high pommel and semi-circular cantle and the seat was richly upholstered with rolls of padding to support the thigh. A long stirrup leather kept the rider's legs almost straight and illustrations show that the reins were held with high hands allowing the weight of the rein to balance contact with the bit.

The airs above the ground, based on natural behaviours in the wild horse, were refined in the manège to show the obedience and beauty of the horse. The skill of the trainer in creating this art form was highly valued in classical dressage and

Fiona Lawrence yard manager, Paula Sells, Carl Hester (see page 98) and at least three dogs in his tack room. There's always a cake on the table, brought in by one of many visitors every day.

court circles and Cavendish enjoyed a reputation as an outstanding horseman.

At the age of 52, Cavendish was obliged by the Civil War to flee to Antwerp but he put his exile to good use by writing the first of his equestrian works, a great folio volume, *A General System of Horsemanship*, published in 1658. It has since run to several translations and editions with various titles, though not appearing in English until 1743. It is a landmark in equestrian literature, produced with the highest quality paper and typography and superbly illustrated with copper plate engravings. It is the most beautiful book I have ever seen and it would be a rare treasure in any bibliophile's library.

Cavendish returned to England after sixteen years in exile but never regained his influence at court. His old pupil, Charles II, gradually abandoned the manège for the racecourse and the combination of this and the effects of the Civil War led to a decline in classical dressage in Britain over the next hundred years.

However, the importance of Cavendish's contributions was acknowledged by continental dressage masters such as Robichon de la Guérinière (1688-1751) whose methods still form the basis of training at the Spanish Riding School in Vienna today. Over the next two hundred years, leading experts in classical dressage all paid tribute to Cavendish. Their works, together with those by Plinzner and Steinbrecht, later contributed to the German Cavalry Manual in 1912 and to the success of German dressage since. Cavendish's influence on modern competitive dressage can be traced through these masters up to the present day, for example, in revisions to the Fédération Equestre Internationale's Definitions of Paces and Movements for dressage in the 1950s.

A measure of the decline in British classical dressage which lasted for nearly a century is in the building of indoor riding houses; between 1660 and 1750 only three new riding houses are known to have been built. During the Georgian revival which followed, at least fourteen were built in the short period 1750 to 1780.

Only the very wealthy could afford to keep and train expensive horses. They built riding houses on their estates which were effectively theatres in which they displayed their skills and wealth to a select audience. A riding house is about the same size or smaller than a modern dressage arena, at least two storeys tall with high windows to avoid distraction and with a viewing gallery usually at first floor level. These were essentially the manèges of the Continent but required a roof and walls to make riding possible in the English climate. Many were elegantly designed by the architects of the day such as Inigo Jones, though were much plainer architecturally than the elaborate continental riding halls of the 18th century such as the present Spanish Riding School in Vienna, rebuilt in 1735.

William Cavendish built riding houses at Welbeck Abbey, Nottinghamshire and Bolsover Castle, Derbyshire, though the very first had been built by his friend, Prince Henry between 1607 and 1609. Riding houses built later during the Georgian revival can be seen at Calke Abbey, Derbyshire, and at Hovingham Hall, North Yorkshire where George III learnt to ride. His teacher, Sir Thomas Worsley (1710–1778) of Hovingham Hall, was a successful breeder of horses and so keen on classical dressage that he designed his riding house as the main entrance to the Hall so that all his visitors had to pass through it. Although it is a very handsome building, the Worsley ladies complained bitterly about the smell from the close proximity of the horses and claimed they were socially ostracised because of it.

It was in these English riding houses, continental manèges and academies that the élite world of aristocracy revealed profound changes in our relationship with the horse. For a wealthy man, his horses were the showcase for his status and prestige. By showing his power to dominate the horse, in effect playing on the romance of taming a 'wild beast', he enhanced his appearance as a leader and warrior. Later, as more enlightened training methods spread, it was seen as more impressive for the rider to be able to perform classical dressage manoeuvres by communing with the horse, his willing servant, and so demonstrating his leadership through skill and harmony. Horsemanship on the battlefield and classical dressage became two disciplines running parallel with one another but were associated through the aristocracy who were often engaged with both the cavalry and the manège.

Modern dressage

Modern competitive dressage grew out of classical dressage and shares the same objectives: obedience, agility and harmony, not forgetting the creation of beauty.

Instead of the emphasis on the superiority of the rider over the horse as it was around the time of Elizabeth I, the focus now is on revealing the natural athleticism and grace of the horse.

Dressage on the Continent thrived and became an Olympic sport in 1912 though only army officers were eligible to compete until 1952 when it was opened to civilian men and women. In Britain dressage remained in the equestrian doldrums until 1961 when the British Horse Society Dressage Group was formed. Jennie Loriston-Clarke won Britain's first dressage medal at the World Championships in 1978 and British Dressage, established in 1998, became the governing body in the UK. Although dressage has always been one of the three phases in eventing, it was not until the last few decades, when Carl Hester and his colleagues showed the world what was possible, that dressage in its own right, especially in Britain, has enjoyed such popularity.

Carl Hester, MBE, was born in 1967 and raised on Sark in the Channel Islands. No cars were permitted on the island and so riding, much of it bareback, was second nature to him as a child. Carl and the British Olympic eventer, Ian Stark, MBE,

Carl Hester with Snowy the cockatoo in the tack room. Carl is holding a double bridle, compulsory for international competitions under FEI rules, with a fixed Weymouth bit.

Carl Hester's stable yard, quiet and relaxed. The day begins with a 6.30am feed and a quiet half hour until 7am, followed by mucking out and grooming. Carl rides the first of many horses at 8.30am. The day's schedule is tight but never hurried. Each horse will have turn-out time in the field and a hack twice a week with a day off on Sundays.

who was brought up in Scotland, have both said how valuable bareback riding has been to them ever since because it establishes an intuitive natural balance. Neither had the benefit of a background with horses yet both seem to possess a double dose of DNA for understanding the equine psyche.

Carl's early training, picked up wherever and whenever possible, for eventing and dressage brought him into contact with some eminent equestrians, such as Jennie Loriston-Clarke and Jane Holderness-Roddam. His first major win in a dazzling career was the Young Riders Dressage Championship at Goodwood in 1985 on a borrowed horse and, at the age of 25, he competed at the Barcelona Olympics making him the youngest rider ever to have represented Britain at the Games. His international career in dressage began by riding for Dr Bechtolsheimer in Gloucestershire for three years and led to 66 national titles, seven European championships, three World Championships and four Olympic Games including two European and Olympic team gold and silver medals to date.

Throughout this time he has brought on young horses and riders at his yard in Gloucestershire which he designed himself. Most striking in its design is his consideration for his staff and horses. The complex is divided into two areas for the benefit of the horses: a quiet relaxation area (stable yard and looseboxes) and a preparation and working area (indoor school, tack room, wash-down stalls and solarium). Maintenance work for the staff is made easier by having a discreetly located but nearby muck heap and a feed store. A covered horse walker which is situated where everyone can keep an eye on it when in use, has a pond in the centre. The pond is home to a variety of busy, noisy waterfowl which no doubt entertains the walking horses and gets them used to noise and movements. A number of other animals enjoy the relaxed environment here; several dogs and more unusually, a mixed flock of ducks, chickens, guinea fowl and even a wild pheasant, parade around free range.

The tack room is light and spacious and adjacent to the saddling up stalls in the indoor school. The green and cream décor is very calming. Most of the wall space not occupied by saddle and bridle racks is filled with photos, competition plaques, cups and rosettes. Leading off from the tack room is a glass-

Carl Hester giving a masterclass at Bolesworth International Show in 2017. Charlotte Dujardin is demonstrating the 'passage' with a six-year-old mare, River Rise Nisa. Both horse and rider (and coach) are in perfect profile.

fronted viewing area with seats for observing work in the school. Not many tack rooms make visitors comfortable by having a round table and kitchen area but this reflects Carl's friendly and relaxed professional style. It is a perfect space for storage, cleaning tack, socialising and talking to clients.

I am greeted by Fiona, yard manager, balancing a pile of cakes as she goes to put on the kettle. As we examine a full-size magnetic Activo-Med rug spread out on the spotless tiled floor, Carl walks in to ask if I would like to meet Snowy, the umbrella cockatoo. Snowy arrives on Carl's arm to spend some time in the tack room. His 'Hello' followed by a piercing shriek makes me jump out of my skin and can certainly be heard by the horses working in the school next door – what a brilliant way to teach them to concentrate through unexpected noises!

Carl has developed the PDS (Professional Dressage Solutions) range of saddles, 'the Carl Hester Collection' made in Walsall (page 8) by GFS (Geoff Fieldhouse, Master Saddler). The objective in their design was to be light (adjustable carbon fibre tree) and flexible enough to give the horse maximum freedom of movement, both in its back (swinging up and down and side to side) and shoulders. Close contact for the horse to feel every nuanced shift of the rider's weight is achieved using panels of flocking and foam and comfort for the rider is helped by a supportive roll in front of the leg. A short wide girth fastens below the saddle flap so that there are no buckles under the rider's leg.

Several different makes of bridles are used such as GFS and those made by Issi Russell (Master Saddler, page 31). Her bridles give extra comfort over the poll, and all the buckles are aligned for neatness. Rolled leather straps are comfortable and show off the horse's head to the best effect.

Carl's own creation of 'Fantastic Elastic' reins are designed to keep a soft contact with the bit and give feedback to the rider on the amount of pressure they are exerting. They consist of a regular leather rein with a parallel piece of elastic sewn in to create a small loop in the leather rein near the bit. When stretched the leather rein comes into play as normal. Until then, first contact between the rider's hands and the bit is through the elastic.

A potential top dressage horse should have several qualities. Good conformation, for example, a well set-on head with room behind the jaw (two fingers) for easy flexion, medium sloping pasterns and strong hocks, good active paces with rhythm and a willing temperament. Most equestrians would put the suitability of a horse's temperament to its job high on the list, whatever the task. For dressage, with its focus on solo work, often in the supercharged atmosphere of indoor arenas, a horse with a calm, confident approach and good concentration would be ideal. KWPN Dutch bred horses are very popular as they have the strength and intelligence to succeed at the top.

Carl has an extraordinary ability to see right into the horse's mind, identify any problems and recognise its potential talent. Bringing out and developing that talent, sometimes obscured by an awkward temperament, is his particular forté.

Many of his successes are with horses who were not particularly straightforward and which most other people would probably have rejected as being unsuitable for the long training needed to reach Grand Prix level. Among the most notable are Escapado, who suffered from severe separation anxiety, often neighing throughout a dressage test, Donnersong, who bucked regularly and Dolendo, who was very slow to mature. Even the now world-famous Olympic winner, Valegro, began as a head-shaker when young and Nip Tuck, all 18hh of him, is a naturally shy and nervous horse. During a dressage convention, while Carl was demonstrating on Nip Tuck, he said, 'If anyone in the audience even drops a sweet paper, this horse will know about it'. All these horses received individual sympathetic management: good intention was rewarded as well as achievement during years of training and they overcame their problems to become international stars. Carl's patience in not hurrying a horse until it is mature enough to understand and his ability to impart confidence can transform it into a willing, expressive athlete. Valegro, outstandingly ridden by Charlotte Dujardin, CBE, at Olympia in 2014 to break the World Freestyle record, is a shining example where equine power and exuberance enthralled the equestrian world.

The aim of dressage is to perform with a relaxed and alert horse that expresses itself through self-carriage, i.e. the ability to hold its profile without depending on the rider's hand or leg for support while still being intently in tune with the rider's aids.

The concept of throughness, in which the impulsion flows from the hocks and hindquarters, over the horse's back and neck and back to the rider's hand without any tension, is a prerequisite for any advanced dressage movements. Throughness is achieved when rhythm, suppleness, contact, impulsion, straightness and collection have been mastered – these are known as the Scales of Training. Straightness, nothing to do with following a straight line, is where both sides of the horse work in a balanced and equal way and this can be developed through lateral exercises such as a shoulder-in or leg-yield, to increase the strength and suppleness of the horse. As the correct muscles become established, manoeuvres become easier and extension and elevation of the paces improve. The rider too can help the horse by increasing his or her core strength and muscular tone. Many riders work out at a gym, take Pilates lessons or practise vaulting which can enhance awareness of the horse's rhythm and balance. A training in dressage is not just an end in itself but also the basis for success in many other equestrian pursuits as it helps the horse to focus its power.

Partnership comes with practice (Charlotte and Valegro competed together for more than eight years) and it all takes a long time. The rider and horse are not the only partnership which matters though. For a competitive horse often on the move between venues, the familiarity of his usual groom, and for Valegro, Alan Davies the travelling head lad, can be very reassuring.

Unlike the management of top dressage horses on the Continent, Carl has always turned out his horses, daily if possible, to allow them to be as relaxed and natural as possible. They all hack out too which gives them variety and experience with meeting various hazards. 'It makes sense', says Carl, 'that if a horse is comfortable, he can give his best'.

It is no surprise that none of the horses in his yard are thought of simply as a commodity. Although selling horses has to be part of the business, very many of Carl's horses return to him for a happy and active retirement when their competitive days are over.

CHAPTER SIXTEEN

Endurance

'A hundred miles on a horse in a day' is a magical phrase that wove itself into Nicki Thorne's young life. As a child growing up in Norfolk, she was fascinated by show jumping on television and loved going to race meetings in Newmarket with her family. Later, as a teenager, she would ride for hours in nearby Thetford Forest, quite unaware of the time or the distance she had covered. Out of curiosity she and a friend joined a local Endurance Riders' group whose members were generally covering distances of thirty to forty kilometres. Measuring the route of her previous rides with the help of a piece of string, a ruler and a map, she realised that the distances she had already covered were comparable.

Over the next few years, Nicki increased the distances until she achieved her first 100-mile (160km) distance and went on to represent Britain in top level endurance competition. The thrill of competing in the countryside is complemented by her interest in Arab racing at home and abroad and, if she were not so tall, she would probably be a flat race jockey as well.

Historical precedents from the Mongol armies to today's long-distance riders (*see* page 150) have shown that the horse is capable of extraordinary endurance. However, neither compare with competitive endurance riding today; the armies of the Steppe usually changed their horses about every forty kilometres (as in the annual Mongol Derby) and long distance riders who travel across continents are not competitive. Horses are tough and found in the hottest and coldest parts of the world in temperatures of −40°C to 60°C. Since they don't hibernate, they maintain a core temperature of about 38°C, a thermoregulation process which requires energy generated by the breakdown of fodder.

Competitive endurance is a controlled race ridden on the same horse throughout the distance. The first horse across the finishing line wins, but only if all the veterinary checks on the horse's condition during the race and on completion are satisfactory. Horses aged five and over may participate in preliminary qualifying rides (classified as Novice) and at six they may enter short and middle distance (Open level). At this age the horses are not racing competitively but are given grades based on a calculation of speed combined with veterinary results, and for long distance and race rides (Advanced) they must be seven years old or over.

Arabian horses have been raced for centuries and in many parts of the world. Lady Wentworth was the daughter of Wilfrid and Anne Blunt who had imported Arabians to England in the late 19th century and established the Crabbet Park Stud.

In her expert view, the Arabian is unrivalled in its stamina as shown by numerous world records. She cites a horse named Crabbet who won over a 310-mile course carrying over seventeen stone (111kg) in 1921, a weight which would be considered excessive today. Crabbet won races over a distance of 100 miles on several occasions and a mare, Ramla, won over 310 miles carrying over fourteen stone (92kg). In races from Newmarket to London in the same era, Arabians consistently won over thoroughbreds.

Endurance riding became an organised sport in the USA in 1955 after Wendell Robie and a group of friends rode across the Sierra Nevada Range on the old Pony Express route. They completed the hundred mile trail in twenty-four hours, continuing during the night, and this trail became known as the Tevis Cup (Western States Trail) competition. It is still one of the most difficult tests of endurance riding in the world due to the mountainous terrain, humidity and extreme temperatures. Approximately half the competitors complete the course.

Endurance competitions are different from racing in another respect. In steeplechasing and flat racing most jockeys, unless 'retained' by a trainer to ride his horses exclusively, don't play a major part in training and may at best have ridden the horse only a few times before the actual race. In a very short time the jockey has to be able to appreciate the horse's temperament and talent and inspire it to give its very best.

In contrast, the endurance rider comes to know his or her horses so thoroughly that every nuance of the horse's performance is familiar. Nicki Thorne is thoroughly tuned in to her horses. At a recent major competition she felt her horse's stride was fractionally less free than usual and, although none of the vets could find anything wrong, she withdrew at the pre-ride vetting before the start. In Nicki's view it was not worth the risk of even the slightest unidentified niggle being aggravated, despite a perfect record of fitness during training. 'It is how the horse is on the day that matters', she emphasised.

Before crossing the start line the rider should look for signs that the horse is settled, such as listening to its breathing (a different sound when working in cold air until fully warmed up), feeling any tension in the back (which may indicate discomfort and reduce its stride length) and noting the look in its eye. This degree of thoroughness is essential for success. It enables the rider to know when to push on and when to ease up so conserving the horse's mental and physical energy, recognising when it has a 'second wind' and when a breather is best.

The competitors' strategies during the race vary; some like to ride alone so they can dictate the pace, others in groups as horses are naturally gregarious. The pace is mainly determined by the terrain and the rider may choose to dismount and walk on difficult stretches. On a hot day, fording rivers is a chance to cool off, the rider throwing a sponge on a string into the water and squeezing it over the horse's neck while still mounted. A plaited mane helps the horse to keep cool.

The horse's condition is monitored before and after the race as well as at intervals throughout the course by a team of vets at each station or 'vet gate'.

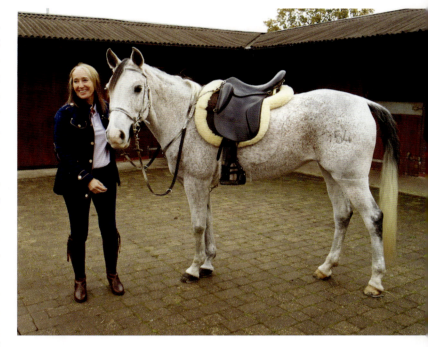

Nicki Thorne with LM Ashirta, a pure bred Arab mare who can trace her pedigree back to stallions imported from the Middle East to the Crabbet Park Stud, Sussex, by Wilfrid and Lady Anne Blunt in 1878.

A number of internationally agreed tests are applied to check that the horse is not lame or dehydrated and has a normal heart rate of 64 beats maximum per minute. The horse is thoroughly checked metabolically and for gait, including muscle tone, lesions, back pain, mucous membrane, and capillary refill. A failed test means immediate elimination of the horse and rider. At each vet check after the tests there is a compulsory rest period for feeding and watering the horse and rider before the clock starts again and the pair continues out on course to the next vet gate. It may be a linear course from place to place, or 'out and return' in a series of loops forming a daisy pattern and starting and finishing each at the venue.

The rider has a personal back-up team (crew) to help at the vet gates which allows time to eat and rest while the team check and look after the horse. In this way families can participate in the race, driving, grooming, catering and helping with all aspects. At the Red Dragon Endurance competition in Builth Wells, mid-Wales, which I watched with Tora Thomas, equine radiologist and endurance judge, it was clear that the back-up teams knew each other well and that they form not just a local network of friends but an international one, all willing to help each other. Out in the beautiful countryside it was like a giant picnic and every rider coming through was enthusiastically encouraged on his or her way.

In most national competitions there are no significant cash prizes for winners, perhaps a bag of feed or a headcollar from sponsors or a tee-shirt. There is usually a special prize for the Best Condition Horse. This can be awarded to a horse that may finish behind the winner but has carried more weight throughout and had good scores at the vet gate checks.

Nicki Thorne's 'Akala Arabians' establishment, named after her successful Arab endurance mare, looks out onto the Norfolk countryside. An attractive landscaped approach to the stable complex leads through an arch to a courtyard surrounded by timber stabling. On the way we pass LR Bold Greyson, one of Nicki's most experienced horses which successfully represented GB twice, grazing in his paddock.

The tack room opens into the courtyard and has plenty of room for visitors to sit and have a cup of tea. Despite the cold weather the room is warm due to

A competitor in the home international Red Dragon contest in mid-Wales with her crew. High-visibility synthetic tack is popular and caged stirrups with trainers are a safe combination.

A custom-built rug dryer in the tack room constructed from hot water pipes diverted from the heating system, popular with people and animals. Both Red the Labrador and Patch the English Setter are trained for falconry.

a heated rug drying rack, designed by Andy Thorne, Nicki's husband, and is particularly appreciated by their dogs who snooze underneath it on a large duvet.

It is clear that correctly fitting, comfortable and robust saddlery is essential for endurance riding. Many competitors use biothane, a synthetic leather substitute which is fairly light and weatherproof – perfect also for its resistance to the buckets of cooling water thrown over the horse at the vet gates.

Searching for a new endurance saddle, Nicki looked at what other riders were using, and seeing that World and European Champion Maria Alvarez was using a Reactor Panel saddle, she contacted Saddle Exchange, the UK manufacturers. Since then she has never looked elsewhere.

The saddle top is of leather, handmade in Walsall, with a conventional wood laminate and steel spring tree. Instead of the usual flocked, air or foam-filled integral panels, flexible panels are made from a sandwich of memory foam and leather which flex with the horse's lateral and medial movement. The panels are attached to the underside of the saddle with sorbothane discs which can be easily positioned on heavy-duty Velcro panels to give superior shock adsorption, a perfect fit for the horse's shape and a correctly balanced saddle. The panels distribute the rider's weight effectively over a larger than normal weight-bearing surface and because they are not joined to each other, they can act independently and in harmony with the movement of the horse's back. This adaptable system means that it is easily adjustable, and the same saddle can be altered to fit a horse both at the beginning and end of its training and fitness programme. The newest model lightweight Reactor Panel monoflap saddle, about 4 kg, is used on youngsters for training.

Stirrups for endurance riding are much larger than usual; they have a spring tread with good grip and a cage can be fitted onto the front so that trainers can be worn safely. Girths are sheepskin lined for comfort. Nicki's bridles are lightweight, silver-coloured synthetic leather, silver stars being her racing colours, and made in Dubai by Horsey. The bits – Neue Schule 'turtle top' bits which suit the conformation of Arab mouths – clip onto the cheek pieces of the bridle and can be quickly detached leaving the headcollar in place. A breast girth and running martingale are used when appropriate.

We leave the tack room to meet a mare, Ashirta, who has come over from the Thorne's joint venture stud in Argentina. Her newly fitted hind shoes have been widened on their outer aspect, correcting her balance and allowing her to stride out freely. Shoeing for endurance riding is a specialist task and care of the horse's feet is crucial. Garrett Ford is a Tevis winner and owner of EasyCare which manufactures synthetic Easyboots and their many variations which are widely used in endurance competitions. Bruising of the horse's sole is easily caused by stony tracks and could result in elimination at the vet checks. Synthetic footwear for horses which provides good traction and protection of the hoof, sole, frog and heels is an expanding science, especially in the

Bold Greyson and Nicki training for Team GB.

US, and has proved to be successful in the most testing conditions. Nicki's horses in the UK are shod traditionally by Laurence Ridgeway, who has attended her endurance horses for many years and works in collaboration with Sion Davies, remedial and consultative farrier.

Endurance horses are elite athletes and their fitness and nutrition is the subject of much scientific research. Dr David Marlin, a distinguished equine physiologist and advisor to British Endurance teams, is Nicki's nutrition consultant and a member of her crew. A recommended diet for endurance horses is a cereal-free diet (low carbohydrate and high fibre) which contains oils and fats to enable them to perform optimally.

Unlike a racehorse, whose maximum effort is measured in minutes, an endurance horse's performance is sustained over many hours (with short breaks) and its diet is based on its energy requirements. Energy in feeds comes from carbohydrates (glucose and glycogen) and fats in various forms. Protein can also be utilised as an energy source but only to a significant extent in times of starvation.

During exercise, energy can be produced by the body in two ways: aerobic (with oxygen) and anaerobic (without oxygen). The type of exercise determines which pathway is likely to be dominant. The limiting factor in endurance is the supply of muscle glycogen.

During intense activity such as sprinting, the anaerobic pathway is activated (even if sufficient oxygen is available). The anaerobic pathway delivers energy very quickly but inefficiently which means it burns up a lot of fuel for the amount of energy produced. A short gallop or steep hill climb will use energy provided mainly by the anaerobic pathway but it will reduce the muscle glycogen store quickly.

By comparison, the aerobic pathway produces energy relatively slowly but very efficiently meaning that a little glycogen goes a long way. Oils and fat in the diet provide the slow-release energy necessary for steady and sustained work.

Diet, training, fitness, time, riding pattern (fast and slow work) and speed all determine the balance between carbohydrate and fat as energy sources and between aerobic and anaerobic metabolism. Finding the right balance of protein, fats and carbohydrates for each individual horse, according to its workload, is one of the challenges of managing competition horses.

An ambition for endurance riders is to develop the closest possible working partnership with the horse, and to share what it can naturally do with ease – covering terrain at a reasonable speed, out in all weathers, facing the challenge of natural obstacles and enjoying spectacular countryside. It is evident from Nicki's enthusiasm that she has found the perfect way to enjoy her horses. Her successes include attaining FEI 'Elite' status, riding as part of Team GB representing Great Britain three times, winning 160km competitions in 2015 and 2016 when she was also national champion on LM 42. She is ranked in the top ten of world endurance rankings (number one in the FEI open world rankings in 2013 and 2014). In 2017 she was elected as Chairman of Endurance GB.

Although one of Nicki's chief joys is riding alone, a team of helpers at home and expert crew who are present at the competition are essential.

ENDURANCE

A Reactor Panel saddle, preferred by Nicki Thorne for its comfort for horse and rider and its adaptability. Blocks in front of and behind the rider's thigh help stability in steep terrain and the sheepskin pad can be changed for a dry one if necessary at each vet gate check.

Among them are Nicki's husband, Andy; identical twins Hannah and Greta Verkerk who, along with Steven Keane, look after the horses; Dr David Marlin, nutritionist; Lee Clark, equine physiotherapist; Pippa Windle-Baker, equine physiotherapist and Laurence Ridgeway, farrier. Being cared for by familiar people also helps to settle the horses when stabled away from home. This management is not just for success on the day but to ensure the longevity of the horse. An unhurried and careful training programme, tailor-made for each horse and prompt attention to the smallest detail have kept Nicki's horses sound and active usually well into their twenties.

Most top endurance horses are pure or part-bred Arabs but any horse can take part if it is physically fit to do so. Arab horses are constitutionally well suited to long-distance riding at speed; their light frame allows them to lose heat well, they have good bone, strong hoofs, stamina and are quick and agile. Most are 15–16hh and there are rules governing the weight they carry during competitions at FEI level (minimum of 75 kg for FEI 3* 160km).

Endurance riding is an international sport, though not yet included in the Olympics. The governing body is the International Federation for Equestrian Sports (FEI) which sets the rules and standards for international competitions and in Britain, endurance competition is regulated by Endurance GB (EGB).

Nicki Thorne on national champion LM 42, competing at Windsor, a picture of fitness and balance.

CHAPTER SEVENTEEN

Cross-country and Three-day Eventing

'There's nothing quite like Badminton!' is often heard from both competitors and spectators. The Badminton Horse Trials are the most famous in the world, rated four star and the heart of British eventing.

They were the brainchild of the 10th Duke of Beaufort who intended them to be a practice opportunity for the British team, which he and Colonel Trevor Horn were helping to manage, for the 1952 Olympics in Helsinki. He lent his home, Badminton House in Gloucestershire, for the first event in 1949 and with the backing of the British Horse Society they continue to be held there.

The present magnificent Badminton House in Gloucestershire is home to the 12th Duke of Beaufort and set in parkland in some of the most beautiful English countryside. A British version of the ancient game of shuttlecock was played in the great hall here in the 1860s, giving its name to the contemporary sport of badminton. The stable yards (*see* Hunting page 127) can accommodate about a hundred horses in permanent looseboxes instead of the temporary stabling found at many three-day events.

Eventing is a combination of three phases: dressage, cross-country jumping and show jumping, ridden on the same horse and within a time limit for each phase. This is one of the great equestrian challenges because of the variety of skills required by both horse and rider.

Calmness and obedience are required from the horse in the dressage phase, which is a set test to show the horse's obedience, suppleness and control of power. It is judged on accuracy and on the horse and rider as an entity (the degree of communication and harmony between them) and it may take years of practice before a horse is ready for advanced competitions.

Stamina, speed and courage are essential for the cross-country course where a top-rated (four star) event of about four miles will have over thirty jumps. It includes water jumps, hedges, gates and rails and galloping stretches as well as tight manoeuvring between fences. The ornamental lake in front of Badminton House is incorporated into the cross-country course with jumps in and out of the water. HRH the Princess Royal and Captain Mark Phillips were two of many competitors who went for an unexpected ducking there. Eventing is a great leveller where success can turn on a split-second of bad luck and sportsmanship is the order of the day. A good cross-country horse will have both obedience and confidence. In the past, this confidence and the evaluation of different types of jumps was often learnt by the horse in the hunting field but top event horses are now too valuable to be exposed to the cut and thrust of the chase.

The final phase of show jumping requires a horse with agility and precision.

The challenge for the rider is to help the horse excel over such a wide range of demands and it takes a very special horse to have both the physical ability and also the right temperament for all three phases.

Forward-going horses who relish the cross-country galloping may find dressage frustrating and boring. Conversely, good dressage horses may find the cross-country jumps too daunting.

The strong bond of respect and affection which can develop between horse and rider over perhaps years of training together is most evident when the riders, breathless and sometimes emotional, are interviewed by the press immediately after the daunting cross-country phase.

There is a formidable roll of honour of both riders and horses in the eventing world. At Badminton, Lucinda Prior-Palmer (Green) won six times and was placed several times. Ginny Elliot (Leng) won three times after breaking her arm in several places, Ian Stark won three times and in 1988 was both first and second. Sir Mark Todd, the oldest winner, won Badminton for the first time in 1980 and for the fourth time in 2011 aged fifty-five, including completing the cross-country course after a stirrup leather broke half-way round. There are many more unsung heroes. Harry Meade was placed third in 2014 having shattered both elbows in a serious fall not long before and Michael Jung won the European Championships despite riding with a broken ankle.

Only two riders have achieved the Rolex Grand Slam (winning at Badminton, Burghley and Kentucky in the same year): Pippa Funnell in 2003 and Michael Jung in 2016. HRH the Princess Royal and her daughter, Zara Phillips, were the European Champions in 1971 and 2005 respectively and both represented Britain in Olympic eventing teams. Their participation did much to bring a wider audience to eventing.

Mary King, MBE, holds the record for being British Eventing Champion four times and having competed in six Olympic Games including the 2012 London Olympics at which Britain won the silver medal. She has won six team gold medals at the World Equestrian Games and European Championships. Besides numerous other successes, she also won Badminton twice, on King William and Star Appeal, and was presented with the solid silver Mitsubishi Trophy. This splendid trophy was designed and made

Mary King at Badminton on Imperial Cavalier wearing an American gag-snaffle with a Flash noseband, a combination which this forward-going horse respected and performed well in.

by Judy Boyt (*see* Introduction) who had modelled it from three Badminton horses to represent the three phases of competition. By coincidence, King William was the horse Judy had used as a model for the show jumper.

It is not only Mary's success at the top international level but her willingness to share her knowledge with aspiring young riders which makes her an outstanding ambassador for British Eventing. Her cheerful and optimistic temperament, combined with great courage and determination, makes her a formidable professional who regularly out-competes younger riders. Her ability to give the horse the confidence to do its very best is a special talent.

Mary and 18-year-old Imperial Cavalier enjoying a good gossip in the yard at home.

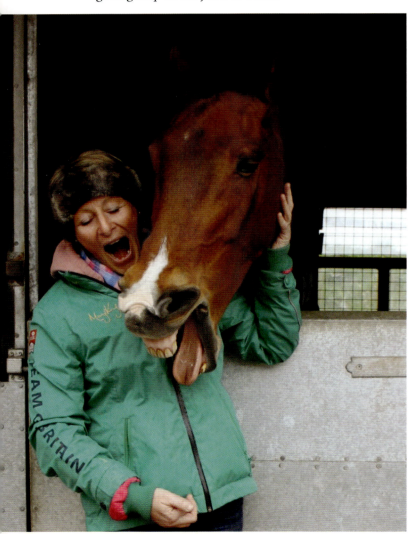

From a non-horsey background, Mary began by learning in the Pony Club, riding anything she could borrow. After competing as a junior on her own pony, she left school at sixteen and spent three rigorous years training in riding and stable management with Sheila Willcox, the only rider ever to have won the Badminton Horse Trials three years running. Willcox was a notoriously humourless taskmaster, once reporting a young Carl Hester to his boss after seeing him hacking along the road with a Jack Russell terrier sitting on the front of his saddle.

Essential to Mary's success is her attention to detail and preparation of all the horses to ensure that competitions are a seamless progression in their development. An important aspect is that the horses should enjoy themselves too and for Mary the reward is the buzz she gets from seeing a young horse (especially one of her own homebreds) reach its potential.

Her yard, which is a combination of old stone barns and timber stables, is on the coast of rural Devon where the hilly country is perfect for getting the horses fit. The premises are immaculate and the eight or so horses stabled there during the eventing season are calm and content in large looseboxes bedded down with chopped plant fibre bedding on rubber matting.

The tack room is a small, whitewashed room opening into a courtyard and filled with saddles and bridles all in show condition. Here are the Barnsby dressage and jumping saddles which Mary has always preferred, including a specially designed jumping model called 'the Extreme Mary King' saddle. Barnsby was the brand name used by Jabez Cliff of Walsall, Royal Warrant holders (1793–2014), who worked with Mary to develop a handmade saddle. Its contours are designed to keep the rider secure and balanced while its carbon-fibre tree makes it lighter in weight and stronger than laminated wood.

The Flair System, in which four sealed (but adjustable) airbags surround a foam insert in the panels, allows maximum comfort, contact and moulding to the horse's back. Mary's saddle is finished with hand topstitching and the seat is trimmed with green leather piping to match her cross-country colours.

CROSS-COUNTRY & THREE-DAY EVENTING

Next door to the tack room is the tea room where we sit and chat and a hen and the cat also hop up onto the bench to join in. An entire wall filled with over twenty magazine front covers of Mary competing shows a long history of success.

The bridle bits Mary uses depend on how forward-going the horse is. All her young horses begin with a snaffle; some may do better later with a vulcanite Pelham or prefer a Waterford snaffle which is more flexible. Occasionally, a gag-snaffle, which exerts pressure on the poll as well as on the corner of the mouth, is used as the safest solution for very strong horses.

Bitless bridles, such as Tim Price's hackamore used for the show jumping at Badminton in 2014, exert pressure principally on the nose instead of the mouth. Ian Stark, famous for his forward-going grey horses, had to resort to a Citation bit to keep control of Murphy Himself. This bit is one of the most severe and was called after the Kentucky Derby winner of that name which was almost unstoppable.

During the eventing season, Mary's tack room is on the road in her Oakley horsebox. International eventing is not the easiest career to juggle with home commitments but in the early days with the help of Mary's mother, Jill, and her husband, David, they made a good family team with their two young children, Emily and Freddie. Mary was the first mother ever to win Badminton in 2000. Emily now events internationally and achieved the top dressage score at Badminton in 2016.

The format of eventing has changed over the years and this has altered the emphasis on the breeding of event horses. When eventing began, a fourth phase of 'roads and tracks' and a steeplechase was also required, originating in the military's emphasis was on testing stamina. The horses then were generally prepared on the hunting field and were usually Irish × Thoroughbred. Standards have risen and competitors have to be very well-qualified and extremely fit to enter four star competitions so, to comply with the World, European and Olympic events, Badminton abandoned this fourth phase for the 'short format' in 2006.

The preferred breeding for top event horses now is Thoroughbred (for speed and courage) or

Mary in her tack room with Barney, the collie. The bridles are correctly 'put up' in Sheila Willcox fashion (throat lash fastened in a figure of eight round the bridle) and the Barnsby saddles are conveniently arranged on wall brackets to the left.

Thoroughbred crossed with continental warm-blood for their tractability and extravagant paces, ideal for dressage. Every year, more ex-racehorses compete successfully in eventing and there is a special prize for the best.

Top eventers such as Mary King and Ian Stark often procure young horses from the Continent as five year olds and then train them here for a few years before competing them (a horse must be at least nine years old before it is eligible for the Olympics). The cost of purchase, training, transport etc. can be met by owners or sponsors upon whom British Eventing depends. Commercial sponsorship of events, such as from Whitbread, Rolex and Mitsubishi Motors, is important, and cross-country fences are often financed by individual companies.

Great improvements in safety have been made particularly with the use of 'airbag' bodywear and frangible pins which hold the top rails of cross-country fences. If the horse hits the rail the pin breaks allowing the rail to drop vertically without rolling forward so that the horse can disentangle its legs safely. Both these safety measures showed their effectiveness in the 2014 Badminton cross-country phase where only 35 of 83 competitors completed the course but no horses or riders were hurt. A new licencing system introduced in 2013 ensures that riders and horses only enter competitions where the standard is within their capability.

Course designers for the cross-country phase are experts in creating courses to test the rider (but never to outwit the horse). It is up to the rider to interpret each fence and communicate to the horse how the obstacle should be negotiated. Badminton winners Ian Stark and Captain Mark Phillips both became course designers but perhaps the most famous was the much decorated Colonel Frank Weldon, winner of Badminton and an Olympic gold medal in 1956 who designed the Badminton courses from 1965 to 1988 and helped to set today's high standard. Hugh Thomas, a former Olympic rider, designed for the next twenty years followed by others, such as Giuseppe Della Chiesa, who continue to challenge the riders with their cross-country courses. Generally, each fence asks a particular question of the horse and rider's skill and some, like the infamous Vicarage Vee, ask several questions at once, such as the competitor's capacity for accuracy, courage and scope.

The design of a successful event and the parity between the three phases, is the subject of much discussion. The penalty marks for each phase are cumulative and the rider with the least number of penalties wins. The tension builds to a peak on the final show jumping phase as the overnight leader is the last to go and the overall winner is often not decided until the last jump is cleared.

Britain has five major three-day events: Badminton, Burghley, Blenheim, Bramham and Blair Castle. All attract huge crowds where country fairs, shopping and camping are enjoyed by all the family. Walking the course is good exercise and just about every breed of dog imaginable can be seen out in the parkland. Throughout the country smaller events, some with all three phases, or cross-country competitions alone, cater for Pony Club members as well as advanced riders and provide the seed corn for top competitors in the future.

Badminton Horse Trials are the most prestigious top level Three-Day Event in the UK and one of only six annual Concours Complet international Four Star events, as classified by the governing body, the Fédération Equestre Internationale (FEI). British Eventing is the governing body which organises events in the UK, as well as schemes to support young riders and is also responsible for selecting and coaching national teams.

If you can be moved to tears by the perfect harmony of horse and rider in a dressage test, experience the adrenalin rush of the speed and unpredictability of the cross-country phase and the nail-biting tension of the show jumping, then eventing is for you – both as a rider and a spectator!

Mary unloading the lockers of her Oakley horsebox after an event. There is space for four horses and living space for six, essential for the many days on the road during the eventing season. Some of the sponsors' logos for Mary and her daughter, Emily, are displayed on the outside.

CHAPTER EIGHTEEN

Farming with Heavy Horses

A ploughman used to walk eleven miles for every acre that he and his two horses ploughed. An acre is the size of a football pitch and would be a day's work. Just imagine doing that on raw winter days wearing a hessian sack over the shoulders to keep off the rain and any nostalgia may quickly fade.

In William the Conqueror's time the design of the plough as we know it today was already recognisable although it was weighty and inefficient. He is credited with introducing destriers (war horses) from Normandy and Flanders as breeding stock to produce the English Great Horse, originally suitable for carrying armoured riders and later used for agricultural work. Interbreeding with continental stock and native British horses from the 11th century onwards eventually resulted in the Suffolk, Shire, Clydesdale and British Percheron breeds.

The introduction of lighter-weight metal ploughs of various designs, particularly those made by Ransome in Ipswich from 1789, allowed horses with their much faster work rate to replace oxen.

Oxen were the commonest draft animals in the Middle Ages because of their superior strength for pulling cumbersome implements but photographs show their use in Sussex and Wales up to the early 20th century, though this was rare by then.

Four-wheeled wagons were depicted in medieval documents such as the beautifully illustrated 14th century Luttrell Psalter, originally from Lincolnshire and now in the British Library. The innovation of the pivoted front axle in the 17th century led to the widespread use of large wagons and British agriculture was powered by horses for the next three hundred years.

The precipitous decline of working horse numbers in the 1950s and 1960s, when the tractors took over, was an alarm call for everyone concerned about their future. In 1983 the Shire Horse Society launched a project, 'History with a Future' for owners, breeders and practitioners of working heavy horses to reinvigorate enthusiasm for the type. Their strategy included the establishment of Heavy Horse Centres open to the public around the country, more ploughing matches, cross-country driving trials and shows.

The efforts of all the breed societies, together with a growing environmental lobby and an increasing awareness of our traditional heritage, have helped to stabilise the four main breeds found today in Britain. The 19th century stud books held by each society are remarkable treasures and document many historical social and political issues affecting horses and their pedigrees. An example of this comes from the Clydesdale Stud book which refers to a Henry VIII Act of Parliament forbidding the sale, exchange or delivery of any English horse to a Scotsman.

King John (1199-1216) had imported a hundred Flemish stallions to improve the British stock and subsequent monarchs had continued to import continental stock. Having raised the standard of English horses through laws regulating the height of breeding stock, this law was designed to prevent Scotland from benefitting from their achievements. The Duke of Hamilton's response was to purchase his own continental stallions and import them direct to Lanarkshire. The law was eventually repealed by James VI of Scotland.

All four breeds are excellent work horses, though the export market, particularly for Shires

A farm iron stack, probably around the 1920s, in Trowbridge, Wiltshire.

and Clydesdales, has encouraged the breeding of longer-legged horses with a smaller girth which are more suitable for exhibiting in the show ring than for working. Their size and impressive presence have always made them favourites for public parades and shows, though for many horselads the only way to reach up to put a collar on a Shire was to stand in the manger.

Percherons originally from the Perche region south-east of Paris, were singled out for preferment as a tough and docile army draught horse in World War I and easy to keep as they are clean-legged like the Suffolks. The Clydesdale, heavily feathered like the Shire, often has a lighter, narrower frame and more elevated paces so is enjoyed as a riding horse as well as for its draught work (*see* Long Riders page 150). Continental breeds such as the Ardennes, Belgian and Comptois are popular in Britain for forestry work.

Collars for agricultural work are of different designs. British heavy horses usually use American or English collars (*see* page 37) or Scottish harness which has characteristic peaked collars and cart saddles. The Scottish collars have a pronounced peak (probably originating from a detachable leather plate or housen which could be raised or lowered to deflect the rain from running down inside the collar), often with an internal moluccan cane frame, and hames which are long and flared. Cart saddles are also peaked and harness hooks, buckles and hames are generally of white metal (nickel) rather than brass, and highly polished.

Trace chains were tumbled in a sack with newspaper as a gentle abrasive for cleaning and leather was rubbed and polished with a smooth wooden 'beetle' to produce a finish like patent leather. Perhaps one of the finest saddle, collar and harness makers of his time was Thomas Prentice (b.1857) of Carluke, Lanarkshire whose excellent craftsmanship and ornamentation is highly valued today.

Blinkers, with plenty of room for the horse's eyelashes, are common on agricultural bridles as are bits which unhook on one side so that the horse may have a midday feed in its place of work without the bridle having to be removed.

Some bridles have a double headpiece; the second one sits on the top of the neck, rather than just behind the ears, and connects to the throat lash or latch. This design makes it impossible for the horse to rub off his bridle against the horse harnessed alongside or on the pole.

Jonathan and Fiona Waterer, Higher Biddacott Farm, Chittlehampton, N. Devon

Horses working in the fields are a regular sight at Jonathan and Fiona Waterer's farm in Devon.

The Waterers' kitchen is probably the busiest I have ever been in; in the space of half an hour about a dozen people had passed in and out, checking details of the day's schedule with Fiona and the phone rang most of the time with Bed and Breakfast bookings, wedding plans and local event details. Outside, the farrier was busy at his work and Jonathan was harnessing two horses and hitching them to a cart.

One horse was a three-year-old, one of a steady stream of horses which arrives at the farm to be introduced to riding, driving and draught work; the other was an older, experienced horse which would act as a tutor. As well as training horses, Jonathan also runs instruction courses for members of the public to learn how to handle agricultural horses. The farm is worked by six Shires who do most of the work in the fields: haymaking, muck spreading, drilling and harvesting wheat for milling and thatching. They take feed in the winter to the herd of Devon cattle as well as doing the daily tip cart run with manure from the stables. Jonathan talks about his work while checking the harness is complete before jumping aboard the cart and disappearing at a brisk trot.

His day begins with a 6.30am feed for the horses and ends after dark. Apart from the farming, he provides horses and carriages for weddings, wagon rides for the public, takes part in TV and film work, and always cuts a dash with the horses in carnivals, shows and parades.

The tack room is at the far end of a traditional stone barn which is divided into several stalls. It has a wide range of harness and saddlery from several countries and for most occasions. A box of red flights sits on the table ready for the next wedding. These are coloured ribbons twisted into a wire stalk and inserted into the horse's mane and tail hair as a decoration. Each heavy horse breed has its own traditional decorations, for example, the four-way diamond row plait with ribbons is characteristic of Scottish Clydesdales. Next to the flights is a coir and leather nosebag which will be the finishing touch for the trade turnout class at the show.

The draught harness used routinely round the farm hangs on the heel post pegs of each horse's stall in the barn. Collars should always be hung up with the narrow end (over the withers) pointing downwards.

At the age of eleven, Jonathan could clean and assemble harness belonging to his father and grandfather – some of which he still uses. After agricultural college he went to Canada to work on a ranch with horses in Alberta, most of them Percherons and Appaloosas.

Working horses are more common across the Atlantic today than in Britain and so innovative farm machinery for them is continually being developed. Communities such as the Amish and the Mennonites have never stopped using horse power but little progress has been made here in that sector since the 1950s.

Jonathan still maintains that link by using Canadian and American designed horse-drawn machinery. In a large shed is an impressive array of dozens of implements closely packed together in a mass of tines, blades, rakes, wheels and gears of apparently whimsical inventions which could have been drawn by Rowland Emett, artist and Punch cartoonist. All these machines are in full working order though and their mechanical simplicity means that they can be easily repaired and will last for decades more.

One of these is a 1951 Nicholson hay turner which is very noisy so is good practice for the horses in their training. Two horses can turn an acre of hay an hour. For heavier work, three or more horses are hitched together which lightens their individual load and means that time is not lost having to stop and

The Waterers' powerhouse! These horses can be hitched up for working singly or all together, depending on the task. Each horse's daily harness is kept on the heel post of its stall and halters and grooming kit are stored handily against the wall.

The Waterers' tack room on their 85-acre Devon farm where heavy horses do most of the work. Western and English riding saddles, spare English and American ploughing collars, a set of chromed hames on a show harness, a bridle with blinkers and a Liverpool driving bit are just a part of this diverse working collection.

rest them. All Jonathan's horses are happy to work in different combinations with each other on a wide variety of implements. American collars (*see* Harness page 34) are used which are lightweight and with an underpad will fit most of the horses. Amish leather traces are an alternative to the chains usually used in Britain.

Jonathan's Canadian team harness breeching is particularly suited to working on the Devon hills where good downhill stopping power is needed. This harness is more effective and comfortable for the horse to stop and reverse on steep terrain than with the British breeching.

The Waterers' farming year, in which Fiona and their two children also take part, is a natural cycle in which the horses are an essential part and which is sympathetic to the environment. The horses supply the muck and nutrients for a good hay and wheat crop which they help to sow and harvest. Their diet of grass in the summer and hay in the winter (supplemented with grain feed) is all within the resources of the farm.

By mowing the weeds and ploughing in the muck heap the result is a sweet pasture on a well-structured soil, natural wildflower meadows, hedges and headlands with diverse habitats. This is traditional farming and the restoration of poor pasture to 'good heart' could take three years or so with benefits down the line for the farmer's son and even grandson to reap. As the saying goes, 'If you look after the land, so the land will look after you'.

For the Waterers, there are many reasons for the enjoyment of having the Shires on their farm but it is particularly for their willingness to do so many different jobs – from taking the tip cart full of muck along a muddy track in the winter to catching everyone's eye with their gleaming presence at the show. The horses thrive on an active and varied working life which is exactly what they were bred for.

Working heavy horses are an extension and a reflection of their handler's skills. The trainer's satisfaction with a good team of horses for a job well done is perhaps even more gratifying than that felt by a tractor driver with his array of automated control systems. As Jonathan says, 'The reason that I employ horses is because I love using them'.

Farming with horses can be a solitary occupation sometimes but is interspersed with regular community gatherings at ploughing matches, harvest, carnivals and shows. Two tales here catch a glimpse of these occasions.

I used to get my heavy horses fit by walking them up onto the mountain with long reins (long lines), which got me fit too. We would often pass by Glyn Jones' sheep farm and he would come out and chat about farming with horses in his youth. He would take the reins from me as we walked, his arms naturally finding that long ago position again after decades of tractor driving.

Glyn told me about a magnificent young Shire horse which was bought by a friend who paid a lot of money for it. The horse was an excellent plough horse but had a quirk; every hour or so he would stop halfway up the furrow and no amount of beating or bribery would persuade him to move. After ten minutes he would willingly set off again as though nothing had happened. It turned out that the horse had been taught to plough by a farmhand who was a smoker. Not wanting his boss to catch him smoking, the lad would stop the horse in mid-furrow and bend over the plough pretending to adjust its land wheel but in fact, lighting his cigarette and smoking it quickly until it was finished. This routine had become so ingrained in the young horse's mind that it kept the habit of stopping every so often for ten minutes for the rest of its working life.

The community and culture of the horsemen was much prized and there were even initiation rituals for the young horselads to instil in them their superiority over other farmhands. There was plenty

Jonathan Waterer with his 1939 Albion reaper and binder, made in England. It has a 5' cut and requires three horses to pull it.

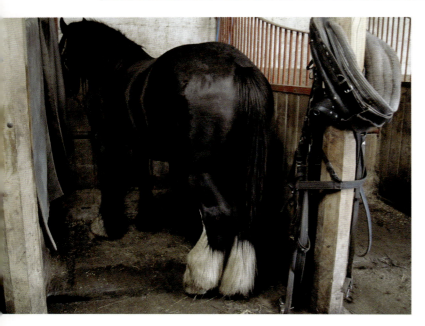

One of the Waterers' Shire horses resting. The American collar, under pad, breeching and traces are handy for the next work shift.

of rivalry between them particularly at ploughing matches and it was not unknown for a competitor to rub ash into the rival horse's coat to spoil its appearance.

Glyn had a story of his own about rivalry. 'I had a neighbour whom I was determined to beat in the ploughing match', he said. 'I used to spy on him when he was practising with his horse and I noticed he put a white hanky in the hedge at the far side of the field directly opposite him and the horse. He taught the horse to aim for the hanky so that the opening furrow they ploughed would be dead straight'. Some weeks later at the competition, Glyn hid behind the far hedge as the neighbour adjusted his plough and set off with his horse. Unnoticed, Glyn quickly hung a white hanky in the hedge well to the right of their starting position and watched while the unsuspecting horse set off towards it. The neighbour, occupied with holding the plough, realised far too late to be able to correct his veering furrow and lost the competition. Glyn and I both had a good laugh and I don't think he ever owned up.

In the heyday of the working horse in East Anglia, the 'sight' for lining up the first furrow was a hazel stick, peeled so that its white pith was easily visible, stuck in the ground at the opposite side of the field. The head horseman would then draw the first furrow and his reputation in both the farm and neighbourhood would depend on its straightness. Such farming news was quickly spread by the grooms who walked the stallions from village to village to serve the mares in the area until at least the First World War. It was common practice for the stallion walkers to hide their fees in the stallion's manger as no thief would dare to steal it.

Horse brasses in Britain have been attached to harness as decorations since the 18th century although in medieval times their brightness was thought to ward off evil spirits. Coloured woollen pompoms, rosettes and bells were also used to dress heavy horses for parades and ploughing matches. In the past, bells hung in a frame mounted on the harness were not just decorative but warned road users of the approach of the horses, especially at night. Large 'journey bells' on a very heavy load pulled by a team warned the Royal Mail coaches to slow down as it was unable to move out of their way, though normally the Mail had priority over all other traffic.

The show harness with brasses was often considered too valuable to be left in the tack room and was kept in the house and polished by the farmer's wife. New brasses are still struck to commemorate special national and local events. The National Horse Brass Society was founded in 1975 by the late Terry Keegan.

Suffolk Punches, Hollesley Bay Colony Stud, Woodbridge, Suffolk coast

This is the oldest stud in the world for the Suffolk (or Suffolk Punch) horse. It began as the Colonial College in 1887 to give an agricultural training to those intending to emigrate – usually the younger sons of landed gentry who would not inherit the family estate. Later, HM Prison Service established adult and young offender prisons there and the stud became part of the prison farm. A pioneering scheme was tried here for a number of years in which supervised young offenders worked with the horses and, it was hoped, would establish a rapport with them which might lead to future employment on release. The results were mixed, the project was discontinued and in 2006, the Suffolk Punch Trust bought the farm to set up a museum and visitor centre.

The tack room is housed in a complex of barns adjoining a whitewashed series of nine stalls. It is a large room with a central table for cleaning tack and a sewing machine for repairs in the corner. Superb sets of urban show harness (the collars lined with leather rather than woollen collarcheck) hang round the walls and more are arranged along the wall facing the stalls. A rota of local enthusiasts keeps the harness in excellent condition. Its next use will be in the show ring at the annual Suffolk Spectacular held in early September and in the ridden and driven Suffolk Endurance Challenge Trophy event.

The Suffolk agricultural collar had wooden hames with hooks for implements such as a harrow but a toggle for the tug chains used with shafts. The cart saddle had the usual leather-covered wooden tree but the pads underneath were laced on with leather thongs. These were much easier to clean or replace than the traditional stitched-in padding. The bridle was often referred to as a 'dutfin' and the belly band for shafts as the 'wanty'.

Tail docking was banned in Britain in 1949. Tails were docked to prevent them from flicking dirt into the driver's face and from getting tangled up in the reins or farm machinery.

Suffolk horses have been part of the British team which took part in the Fish Race, a biennial race from Boulogne to Paris re-enacting the competition to be first to reach the lucrative Parisian market with

The Suffolk Punch Trust tack room at Hollesley Bay. This traditional English harness is so cherished that local volunteers come in regularly to clean it.

fresh fish and command the highest price. Teams of heavy horses from Britain and several European countries took part in the relay race which continued through the night, each carrying a token crate of fish.

Apart from the horses at Hollesley Bay, a large collection of farm carts and wagons complements their agricultural history.

A wagon is full of information – from a glance at its design it is possible to tell the date it was made, which county it came from and a fair bit about the local agriculture there. The decorations such as spindle work and chamfering can often identify the maker and these techniques also reduce the overall weight. The variations of design and the skills passed down reflect a pride in the locality and a way of life not just for the maker but for the wheelwright, blacksmith and horseman all closely associated with its manufacture and use.

It was also intended to last for several generations, in fact to have to repair a cart or wagon in its early years was considered to be a shameful task – this was before the concept of built-in obsolescence, disposability and the acceptance of waste which, in the view of these craftsmen, were introduced with the motor car. Before about 1900, most wagons (county wagons) were individually made in a wheelwright's shop. Later, wagon-making businesses became common in towns where larger numbers of plainer vehicles were produced.

Photographs of special occasions are often found on old postcards and in books and show both the cart and the horse's harness. They may record a children's outing, a wedding or taking the hop-pickers to the station at the end of the week, so the details of their lives have survived. Few pictures were ever taken of the horses doing their daily job of work on the farm.

Records have been kept by all the breed societies but the Suffolk Horse Society's Stud Book, Volume One, compiled by Herman Biddell and published in 1880 is thought to be the finest pedigree register for any type of stock animal in Britain. Illustrated and leather-bound it has over seven hundred pages of the details of individual Suffolk horses and is a valuable historical resource for the breed. Every Suffolk can be traced back in a direct male line to four main foundation stallions, the most famous being Mr Crisp's Horse, foaled in Ufford in the 1760s.

The rich coppery colour, characteristic of Suffolks, is divided into seven different shades of 'chesnut'. Their round and powerful profile shows how suitable they are to work the heavy soils in the west and south of Suffolk and their clean-legged, lack of feather makes them easy to care for – fewer burrs and less clay stick to their legs, unlike the Shire and Clydesdale. They are noted for their longevity and ability to thrive on meagre rations.

The large number of Suffolk stallions exported to the US, Canada & Russia (for army horses) in 1912-13 and the mechanisation of arable farming since then have both depleted the breed numbers. In the 1960s when the large arable estates which had used teams of Suffolks invested in tractors, the

Part of a collection of horse brasses, show harness decorations and lawn boots belonging to the late Mark Roberts, Denbighshire. The lawn boots were worn by horses when pulling a roller on lawns and bowling greens.

Bill Smith, ploughing at Bucklesham with Violet and Teazle in October 2015. The Suffolk Horse Society works to promote and preserve the breed through shows and events such as this ploughing match.

horses disappeared almost overnight. Today there are several Suffolks working in the forestry industry.

The future of the Suffolk is considered to be at risk and foal numbers are low and variable. Due to judicious breeding the health of the breeding stock is remarkably good; a tendency to side bone (ossification of cartilage in the foot) and brittle feet is rarely a problem now and foot classes at shows make sure that the standard of breeding stock is maintained.

The Rare Breeds Survival Trust and the Horserace Betting Levy Board give some monitoring and financial support to the breed societies. The Suffolk Horse Society and Ipswich Record Office both hold archive material and memorabilia of the history of the breed.

Some of the finest books about East Anglian farming were written by George Ewart Evans. He collected aural histories from retired farm workers who lived during the cusp of horse work and the mechanisation of farming which preserve in detail the working era of the majestic Suffolk Punch.

Collections of equipment related to farming with heavy horses can be found at the Museum of English Rural Life (MERL) in Reading and the Weald and Downland Living Museum in West Sussex as well as in many other museums.

There is no official government funding or National Studs for our native heavy breeds, Shires, Suffolks, or Clydesdales, as there are in Europe. The Working Horse Trust, Tunbridge Wells, Kent, is a charity which breeds, works and shows these magnificent horses to keep them in the public eye in order to prevent their decline.

CHAPTER NINETEEN

Harness Racing

In the bracing salty breeze off the Irish Sea, the harness racing at the Tir Prince Raceway at Towyn, North Wales, is in full swing. Eight pacing horses glide round the oval track at speeds approaching thirty miles an hour, each with a driver in a fragile sulky wearing coloured silks, helmet and goggles: a modern charioteer.

It is impossible to see at this speed the order in which the horses use their legs and I can only tell whether the horses are 'pacing' or 'trotting' by the harness they are wearing. A pacing horse moves its front and back legs forwards and backwards on the same side simultaneously, similar to camels and giraffes. It wears light loops (hopples) around the top of the legs which are connected to each other on the same side and encourage the horse to maintain its pacing rhythm. Normal trotting (also known as a square gait) is the usual diagonal sequence where the left fore and right hind move forward and vice versa in unison and hopples are not worn.

Small pacing horses in Chaucerian times, known as palfreys, were preferred by ladies because their gait was more comfortable to sit to. Pacing (sometimes known as ambling) is said to be about three seconds faster over a mile than trotting.

The object in competitive pacing is to race the horse as smoothly as possible (except for overtaking) at a consistent speed round the track, usually two laps of a half-mile oval track. Harness races are started on the move – quite different from flat racing (starting stalls) and National Hunt racing (a general line up from a near standing start). The horses and their drivers assemble behind the starting gate: a race car on which two broad metal wings are mounted, extending either side of the vehicle across the track. Having assumed their appropriate handicap positions, the horses follow the moving starting gate car towards the start post. As the post is passed, the wings retract and the car speeds ahead to leave the track. Breaking pace (trot to canter) of over fifteen strides during the race is penalised by elimination and there are strict rules for clearance when overtaking.

It was not until the 1870s that the horse's gait was more clearly understood. The respected American breeder of trotting horses, Leland Stanford (founder of Stanford University, California), paid the English photographer Eadweard Muybridge (1830-1904) to settle a bet by showing whether or not all four of a racehorse's hoofs momentarily left the ground at the same time during a gallop. As well as the racehorse, Stanford included his pacing horse Occident in the experiment. By using a dozen cameras triggered by threads along a track, Muybridge showed that the horses' legs briefly all left the ground together during both the gallop and the trot and Stanford won his bet. Muybridge's photographs immediately influenced the work of equestrian artists who no longer depicted speed by showing horses unnaturally spread-eagled.

One of the best subsequent painters of movement and muscular power was Lowes Luard (1872-1944) in his studies of draught horses and racehorses.

The superb hard track at Towyn on the North Wales coast was the vision of the late Billy Williams. Billy's first introduction to harness racing was in Prestatyn where he had been invited as a guest to a

The home straight in a mile pacing race at Dunstall Park, Wolverhampton, 2017. The extended lateral movement of the fore and hindlegs is clearly seen.

race meeting in the 1960s. He was so taken with the sport that he came home that day with a racehorse and all the equipment in exchange for his car. He went on to achieve great success in breeding, training and driving his own horses to win national championships. The Prestatyn track closed and by 1989 he had decided to extend his passion to building a new American-style, half-mile circuit in nearby Towyn with a banked track of washed sand and limestone powder, complete with floodlights and adjacent stabling. Just after the track had been built in February 1990, storms caused extensive flooding and completely devastated the new raceway. Undeterred, Billy found £50,000 and rebuilt the circuit, opening on 22nd May, only three weeks later than originally planned – an extraordinary achievement.

A second Welsh harness racing landmark, not far from Towyn near the small cathedral city of St Asaph, is the Saunders Stud founded in the 1960s by John Saunders Jones, known as J.S. Just like Billy Williams, J.S. happened to drive past a race meeting at the old Prestatyn Raceway and was smitten by what he saw. With an ambition to breed his own horse, everything fell into place when, as a businessman, he was asked to provide quarantine facilities for Standardbred stallions which were being shipped from America to Australia via the UK. As payment he asked for two free services from the pick of the stallions. Astutely selecting the best mares he could find in New Zealand, Europe and America, he created a breeding centre in North Wales which became renowned worldwide.

Jim McInally (Vice-Chairman of the British Harness Racing Club) and Mandy Stanley took over the reins of the Saunders stud in 2003, breeding, training, racing and selling Standardbreds. The stud is surrounded by rich green pastures in the River Elwy valley, perfect for raising youngstock. The yard feels like a small piece of the New World with an American-style barn which was one of the first of its kind in North Wales.

The tack room is a whitewashed stall in the barn next to a washing-down wet room for the horses. A wide central aisle divides two rows of looseboxes. The atmosphere is cool, shady and quiet, the horses are settled and relaxed.

In the tack room an array of unfamiliar harness hangs next to photographs of pacing races, the horses flying round the track. Almost all the tack is imported from America (Big Dee, Ohio) where harness racing has been popular since the Civil War. Hopples are light loops which fit loosely round the top of each fore leg and hindleg and are suspended by hanger straps to keep them above the knee and hock.

The loops on the nearside are joined as a pair, as are those on the offside. Their function is to encourage the legs on each side to move in unison and to act as repetitive reminders.

A horse fully equipped for both training and pacing on a race day seems to wear a lot of tack, though every strap is there to guide rather than to restrain and helps the horse to maintain its rhythm and balance at speed.

The harness is made from biothane instead of leather. It is lightweight, strong, maintenance-free, relatively inexpensive and comes in a range of colours. The hopples are soft enough to only require a light dressing of oil to prevent them from rubbing the horse, whereas graphite was rubbed into the old leather hopples as a lubricant – a messy procedure and unsightly on a grey horse.

The head position is all-important for optimal performance and the balance of the horse's body. Teaching aids for the horse consist of a number of light plastic moulds which can be arranged around the jaw during training to correct the head position. For a horse which 'hangs' to one side during a race, a light pole can be clipped between the saddle and the bridle cheekpiece to keep its head straight.

An overcheck rein to keep the head up runs from a separate bit in the mouth over the top of the head back to the saddle. A number of designs of goggles can be attached to the bridle to prevent the horse from seeing backwards and being distracted. Sheepskin nosebands can stop horses from looking down and jumping shadows during a race. Reins are of different designs such as ladder reins which are connected by a series of transverse straps and easy to hold with one hand, leaving the other free to carry a light whip.

The American barn at Jim McInally and Mandy Stanley's Saunders Stud, Denbighshire. The exercise tack for each horse is raised up on a pulley outside each loosebox. The wheels of a sulky with shafts uplifted can be seen on the left.

The sulky is a two-wheeled light vehicle in which a single driver sits leaning back in the 'ten to two' position. The name is thought to have originated from the fact that the driver is always solo, never in a pair, and presumed to be 'sulking'.

The weight of the driver is not critical since it is borne by the sulky wheels and not by the horse. Jim McInally claims to be able to date any photograph of racing by the design of the sulky, for example, contemporary sulky shafts are attached with a simple pin and bracket to the saddle instead of with leather straps.

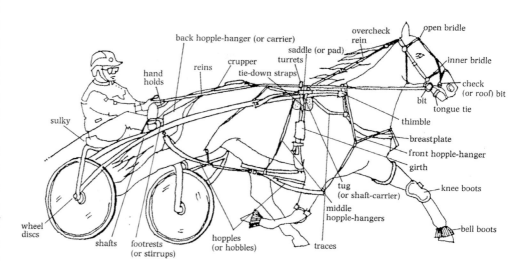

Basic harness used on a Standard bred pacer (tongue tie and boots optional).

Harness racing in the UK is affiliated to the regulatory body, the British Harness Racing Club (BHRC). Competitors have to take an oral and practical driving exam to receive a licence to race and are initially restricted to six qualifying races. Upgrades to higher standards, including qualifying to drive at international meetings, are based on experience and performance.

Ridden trotting and pacing races, using a general purpose saddle, are being re-introduced at some meetings and competitors have to qualify under BHRC rules.

Trotting and pacing races are held on both hard and grass tracks. Hard tracks are located round the country, for example in Wolverhampton, York, Wreningham in Norfolk and Corbiewood in Scotland, and each has its own character. Those who prefer the more relaxed racing on grass have formed the Grass Track Association.

Musselburgh near Edinburgh is a popular grass track, the only one to run clockwise. The Tregaron grass track in Ceredigion, south-east of Aberystwyth, staged its first Welsh Classic race in 1984 and with its substantial prize money, television coverage and fine facilities has a famously lively and friendly atmosphere. Ceredigion is an area in which unaffiliated racing has also been popular and includes galloping as well as harness races, though it does not have the advantage of the organisational and financial support of the BHRC.

Harness racing can count many families in whom three generations have actively participated in all aspects of the sport though the strong spirit of camaraderie doesn't diminish the competitiveness.

The origin of harness racing and the development of the Standardbred horse (a fast, light driving horse) was in the north-eastern states of America. In the 19th century, New England culture was strict Presbyterianism and horse racing was banned as being un-Christian. However, good roads for horse carriages favoured a population of light driving horses and impromptu races were held at fairs. The Courts decided that if the horses kept their gait to trotting, they could not be presumed to be racing. Their reasoning was that if racing competitively they would be going their fastest i.e. galloping. Therefore, if they were not galloping they couldn't be racing. Trotting and pacing became a national sport in America and spread across the world. Today pacing (although banned in Europe) is generally more popular than trotting.

The Standardbred horse can be traced back to the thoroughbred Hambletonian, the great-grandson of Messenger who was imported in 1788 to Philadelphia from England. The progeny from this line showed great ability in trotting and pacing and

Red biothane pacing harness worn by Ystwyth Valley driven by Lee Morris.
Note the light pole strapped between the bridle and saddle designed to stop the horse's head hanging to the right and destabilising his balance.

were subsequently interbred with Hackneys, Arabs and many other breeds to produce the Standardbred. The ideal action for both a pacer and trotter is free and extensive with a long reaching stride, rather than the high knee action seen in pure Hackneys.

To qualify for registration in the Standardbred Stud Book, a horse must 'meet the standard' by trotting or pacing a mile within two minutes thirty seconds. The current UK pacing record is a mile in one minute fifty-six seconds. However in America and Canada where the breeding and racing of pacers is in a league of its own, one of the fastest race mile times was 1:46.4, recorded in 2008 by the legendary pacer Somebeachsomewhere. He retired having won twenty of his twenty-one starts and over three million dollars in prize money.

The horse is probably the only quadruped to have a choice of gait at the trot. The advantage of this in the wild is obscure but recent research is beginning to reveal how this curious process may work. Achieving a correct pacing gait comes from a mixture of breeding and training in which the brain reconfigures its coordination of movement. The extent to which the equine brain is already attuned to pacing is unknown but breeders have noted that when a pacing horse and a trotter breed, the offspring is likely to be a natural pacer suggesting that it may be a dominant trait.

THE TACK ROOM

It is known that a very high proportion of Icelandic ponies are pacers. This trait was no doubt encouraged through selective breeding as it was the most efficient way to travel long distances over terrain where there were few if any roads.

In 2016, a European collaboration of geneticists, archaeologists and analysts identified a gene which plays an important role in the development of limb coordination and has shown that it probably originated in medieval England. Looking at ancient DNA data from ninety horses, the scientists concluded that the pacing gene first appeared in two English horses in York who lived around 850-900 AD. It was absent in all continental and Scandinavian horses at or before that time. However, it was carried by ten of thirteen ponies who lived in Iceland in the 9th to the 11th century. It is suggested that the Vikings, who were already established in York, took pacing horses with them to settle Iceland in the late 9th century, there being no indigenous horses there.

The pacing gene gradually spread out from England and Iceland to the rest of the world along the Viking trade routes. There are 15th and 16th century records in the archives of the city of Modena in Italy recording the import of Irish horses which are thought to have been pacers from their description. They were well bred (probably with some Arab blood), highly valued and known as 'hobbies' – a 'hobby horse' was one used for hawking. According to Dr Samuel Johnson's Dictionary (1755), a hobby is a species of hawk (in France, a hobereau is a bird of prey and a hobin is an 'easy-paced' horse).

Now that the pacer with its comfortable special gait as a mode of transport has been overtaken by progress, this interesting trait continues to be conserved by the British harness racing fraternity.

CHAPTER TWENTY

Hunting

The Duke of Beaufort and Fitzwilliam (Milton) Hunts

The hunting on horseback of various wild animals has a long history in Britain and is colourfully depicted in medieval tapestries, paintings and documents. Fox hunting is the most deeply ingrained in our culture and the animal welfare and social issues it raises are the source of continuing debate.

The history of the hunt, its aims and organisation are outlined here and provide a background for visits to two outstanding hunting tack rooms and to the Fitzwilliam Hunt kennels. The Hunting with Dogs Act 2004 has had a profound impact on the fox and the countryside and its success or otherwise is considered here in the spirit of its current controversy.

A Brief History of Hunting

Hunting has been controversial since the 11th century when William the Conqueror first hollered 'Ty a hillaut' in the English countryside. This was a French medieval rallying call used by stag hunters from which our 'Tally ho' was derived. Their patron saint, St Hubert, Bishop of Liège in 708 AD, advocated 'a virtuous hunt' in which ethical principles must be observed, for example, an animal with young should not be killed.

Even before the Normans had arrived with their hounds and horses, King Canute had designated royal forests in Britain for his private stag hunting. The Conqueror, however, vastly extended the royal forest areas and increased the severity of the Forest Laws causing great animosity and hardship among the population. Local peasants had to apply for special licences to make charcoal, collect firewood and to graze their pigs (pannage) in the forest and a poacher caught red-handed could expect harsh punishment.

Following the Norman succession, the Tudor and Stuart monarchs, including Queen Elizabeth I, were keen participants in stag and hare hunting to the accompaniment of pipes and horns. The privileged few who hunted in this period did so not just for sport and exercise but to display their status at court or to practise their skills for war. By the end of the 17th century the Civil War had left the royal forests and the deer population in decline, the Forest Laws were eased and the hunting of the remaining deer, as well as hares and foxes became increasingly widespread involving landowners and communities rather than being limited to the aristocracy.

The popularity of organised fox hunting led to the formation of local hunts who established themselves with boundaries and their own packs of hounds. The Brocklesby Hunt in Lincolnshire has the oldest private pack in the country and breeding records are complete from 1746, while the Cottesmore claims to have hunted foxes since 1666. Several hunting packs today, descendants of these 18th century hounds, continue to be kennelled on their original estates.

The fox gave excellent sport by running in relatively straight lines compared with the hare and so provided a good gallop for the followers on horseback. Farmers, upon whom the hunt depended

Side saddle safety attachment for the stirrup leather. Stamped 'E.W. Mayhew Saddler to the King'. This saddle, which is still in use, is likely to be pre-1940 when the company was taken over by Champion & Wilton, London.

for permission to ride over their land, encouraged hunting as a way of keeping the fox population in check.

The Enclosure Acts beginning in 1774, were initially greeted with consternation by the hunting fraternity (and outrage by the peasantry) until the new fences were seen as a benefit by providing good jumping opportunities. The Enclosure Act and the Industrial Revolution encouraged country people to migrate to the towns, lifestyles changed radically in the 19th and 20th centuries and debate opened up on many facets of British life, including hunting.

The Game Laws in the 1800s partially protected deer, hares and game birds but foxes were considered to be vermin and so could be hunted by anyone. However, it is said that anti-hunt protesters of the period used to drag a rotten herring across the line of scent to confuse the hounds and this may be the origin of the phrase 'a red herring'.

A wave of cultural awakening, influenced by the wealthy and adventurous who made The Grand Tour of the Continent by coach and horses, introduced new architectural styles which can be seen today in the Palladian and Georgian great country houses and fine stable yards. In this period of prosperity and patronage, many of the landed gentry bred their own packs of hounds.

Fox hunting flourished in the mid-18th and 19th century, becoming more socially inclusive and developing its own traditions, art and literature. The heyday of modern hunting was probably before 1914 and between the two world wars when the countryside was less interrupted by roads and urban development.

Siegfried Sassoon's *Memoirs of a Foxhunting Man* and fictional characters, such as Jorrocks by R.S. Surtees, have left a rich and often humorous literary record of this time. Horses were ridden to meets or were harnessed as a tandem pair to a carriage. On arrival the leader would be saddled and go hunting while the wheeler took the carriage home. The railways enabled horses and riders to travel long distances to and from meets, arriving fresh and relieved of the long hack home at the end of the day. A second horse was often brought by the groom so the owner could continue hunting in the afternoon while the groom took the tired steed back to the stables.

The number and quality of equine paintings in the 18th and 19th centuries attest to the popularity of horses for sport in the culture of this time. They are also incidentally helpful in showing us the details of tack which is no longer commonly used, for example: the cavalry saddle (Munnings, 1878-1959) and coaching harness (Cooper Henderson, 1803-1877 and Benjamin Herring, 1830-1871). Hunting tack itself has changed little over the decades as shown in paintings by 'Snaffles' (1884-1967), Lionel Edwards (1878-1966) and Cecil Aldin (1870-1935) who was Master of the South Berkshire Hunt though few use a double bridle these days.

In the 21st century the population is now more urban than rural, town and country have become more remote from each other and hunting has become increasingly controversial. The RSPCA and the League Against Cruel Sports (founded in 1837 and 1923 respectively) lent their support to the anti-hunting debate which culminated in the Hunting with Dogs Act of 2004. The Act, which passed into law on 18th February, 2005, banned hunting with a pack of hounds in England and Wales and substantially altered the way in which these hunts had operated for the last three hundred years. Scotland had passed a broadly similar law in 2002. Northern Ireland has resisted a ban.

The law now allows a pack of hounds to follow an artificial scent trail previously laid by hunt staff or to flush out a fox either to waiting guns or to a trained bird of prey, but they must not give chase.

Hunting tack

Hunting tack is designed to stand up to robust use in all weathers and should be made of properly tanned leather which is heavier and stronger than the type of tack used for showing classes today. A good stirrup leather for hunting would be at least 1" wide and made of buffalo hide. Each horse is fitted with the type of saddle and bridle which best suits its individual character: appropriate for fast work and jumping as well as comfortable for hours of riding at a time.

The hazardous nature of riding to hounds inevitably results in a number of falls, usually during jumping. To address this, a number of quick release mechanisms for the stirrup leathers and the stirrups themselves were patented. This was particularly true for side saddles, so that the rider was freed if the horse fell. These safety measures are just as important today not just for hunting but for all types of riding, and innovative designs of stirrup and stirrup leather attachments continue to be developed (*see* Lorinery page 41).

Hunting Tack Rooms

The relative stability and affluence in 18th and 19th century Britain enabled the aristocracy and landowners to build stately homes, many of which have become part of our heritage. These grand houses were an expression of wealth and power and their stable yards were also built to impress. For example, the magnificent tack room at Manderston, Berwickshire, with its shining racks of saddles, marble floor and mahogany fittings was also a show piece for visitors to the estate.

His Grace the Duke of Beaufort (see page 127). Henry Somerset, 10th Duke of Beaufort, Master of the Beaufort Hunt, riding one of his favourite hunters, Oxo. The sporting artist, Lionel Edwards (1878-1966) hunted with many different packs, often sketching from the saddle.

The most splendid tack rooms were built to house the tack used for hunting and carriage driving. Some of these still exist and offer an insight into our social history as hunting is the oldest equestrian sport in Britain. It has continued through the age of the carriage and the cavalry horse and both harness and military saddles can be found in some older tack rooms alongside hunting saddlery. Army officers often spent much of their off-duty time hunting and keeping up their equestrian skills. As cars (known as 'horseless carriages' in the 1890s) began to replace horses, the carriage houses were converted to garages, the harness sold or discarded and the tack room remained primarily for hunting use.

The stable yards on large estates also provided the living quarters for a hierarchy of staff from the head coachman to the stable lads. Rows of looseboxes, carriage houses, tack and feed rooms, the smithy and fire station with horse-drawn water bowser and leather buckets could extend the stables complex to the size of a small village, as at Calke Abbey in Derbyshire. Many of these old tack rooms have a fireplace or were warmed by one from the adjoining room of the groom or coachman in charge. In these the tang of wood smoke in the beams is still noticeable and mixes richly with the smell of leather. In affluent yards, hunting tack rooms with girth dryers and stretchers, boot and whip stands and saddle horses on cast-iron wheels would be in regular use though they are rarely seen today.

Before the use of stainless steel, the burnished steel alloy used for bridle bits and stirrups was prone to rusting, hence the positioning of a traditional glass-fronted, baize-lined cabinet above the fireplace in the tack room which was designed to keep them dry. Saddle brackets were made of wood or iron covered with fabric to protect the lining from iron mould. Chests for storing rugs were made of cedar wood to discourage moths and American pitch pine was used for saddle horses and racks because the high resin content deters woodworm.

There was plenty of help to clean and repair the tack in those days and perhaps it is due to the care given then that very old tack, such as the side saddle, has not rotted away but with some restoration work can still be used eighty or more years later. This old tack is lovely to handle and provides great continuity with the past.

A tack room of age and distinction is often known as the 'saddle room', perhaps to distinguish it from rooms where the carriage harness was kept (in the harness room). Two such saddle rooms, from Badminton House and Milton Hall, are described below, both of which preserve the hunting tradition in style.

The Saddle Room at Badminton House, Gloucestershire

The Duke of Beaufort's Hunt at Badminton House is one of the largest and most distinguished in England. The House was substantially altered by William Kent in 1745, who added its distinctive twin cupolas, and it stands in a glorious setting looking out over the deer park and the Gloucestershire countryside.

The Dukes of Beaufort have continued a strong foxhunting tradition since 1762, the 10th Duke of Beaufort (1900-1984) known as 'Master' being the most renowned, not just for his mastership of the Beaufort Hunt for sixty years, but as a skilled and influential hound breeder. He also founded Badminton Horse Trials and was Master of the Horse to HM The Queen, as his grandfather had been to Queen Victoria. His cousin, David Somerset became the 11th Duke of Beaufort and succeeded him as the senior master until his own death in 2017. Captain Ian Farquhar, the Queen Mother's equerry, was appointed joint master and huntsman of the Beaufort in 1985. In 2012 he handed the horn to Tony Holdsworth and continues in his 45th season as a Master of Foxhounds.

The stable yards, expanded in the Victorian era by John D. Tait with additions in Edwardian times, can accommodate a hundred horses. Every May it is full for the Badminton International Horse Trials, now in their 69th year, the hunt horses having been turned out to pasture in April after the winter hunting season. Except for the introduction of electricity, little has changed since 1876, the date inscribed on a lead rain hopper.

Quadrangles of gravel are surrounded by stable buildings of buff Badminton render. Royal blue paintwork reflects the estate colours, though the hunt staff wear the Beaufort family livery of green. Atop the cupola over the stable yard arch swings a fine weathervane. It is the silhouette of a fox, running with the wind, as is their habit. This is a place where

The Badminton House saddle room. A special feature of the Badminton saddle room is the heating system which provides a comfortable ambient temperature and prevents mildew on the leather. The 4" diameter iron heating pipes run the length of the wooden slatted saddle horses and through the floor-to-ceiling airing cupboard for horse rugs. A row of a dozen looseboxes can be seen beyond.

every detail matters – from the length of a horse's tail (2" above the chestnut) to the recessed taps in the stable aisles and folding saddle racks outside each loosebox door.

Before he retired, I spent a day with the late Brian Higham, stud groom to the Dukes of Beaufort for over fifty years. On a typical working day, the ring of hoofs echoed across the courtyard as we watched the huntsman's horse being trotted up and down to check for soundness after a day's autumn hunting. Most of the hunt horses are part thoroughbred and part Irish draught, breeding which typically produces excellent quality of bone and a lively spirit.

Brian expertly assessed that all was well, save for a cut on the heel probably from a flint stone. The horse was led away to be hosed down and dried while the saddle and bridle were taken to the outer tack room, the mud was washed off and the leather gently dried and oiled. The huntsman's saddle had a leather case strapped on either side of the pommel, one for the hunting horn and one for a pair of wire cutters. Behind the saddle hung spare sets of hound 'couples' for use with the young hounds just entered this season.

At the heart of the stable yard is a magnificent saddle room, horse washing area, tack cleaning room and Brian's office from where he presided over the yard. Here all his detailed planning of the day took place. It may have involved a surprise visit from a member of the Royal Family, invitations to judge at shows, the care and maintenance for about 25 resident hunt horses and requests to give an after-dinner speech. Brian had a store of good stories going back to Sheila Willcox's unequalled three consecutive Badminton wins and to winters in the 1950s when

The ground floor of Milton Hall's Georgian saddle room showing the main heating pipe (left) rising up to the drying room above. The movable saddle horse with storage cupboards is a superior piece of furniture not often seen now. On the far windowsill is a brace of leather pistol sleeves dated 1940, possibly for stag hunting.

the Beaufort hunted six days a week.

A daily record is kept in a Victorian lectern desk which still contains the ledgers for the last hundred years or so where the wages of some forty staff are itemised. A glass case over the fireplace displays rows of copper hunting horns some with original ivory mouthpieces, an antique paraffin singeing lamp and drenching bits.

The saddle room is spacious with twenty foot high ceilings and long windows. It is on such a grand scale that I was momentarily lost for words. The wooden floor, panelling and shining leather give the room a soft acoustic, no doubt appreciated by the horses stabled beyond. The walls are covered with rows of saddles and bridles, the topmost being reached with a ladder. Most are used for hunting but there are also cart saddles and collars which once belonged to Percheron heavy horses working on the estate. The fault lines of time are also emphasised by Army Universal Pattern saddles alongside headcollars with padded headpieces for air travel as the British Eventing Team trains at Badminton.

Dozens of double bridles line the walls. Leather girths rather than synthetic are considered to be correct and no saddlecloths or numnahs are used.

Saddles are routinely returned to the saddler for reflocking. The refurbished saddle is then used for a minimum of three weeks to 'bed into' the horse's particular back shape before being returned to the saddler for adjustment and to have its final linen lining sewn in. Some leather lined saddles are also found here and although these are hard wearing do not absorb the sweat so effectively.

The central heating system is designed to run through the saddle horse, up through the storage cupboards for rugs and to outer rooms used for washing the tack after hunting. This excellent design allows the linen lining of the saddles to dry out and no doubt a warm saddle is also appreciated by the horses on a cold morning.

The Saddle Room at Milton Hall, Peterborough, Cambridgeshire

The Milton Hall saddle room stands proudly within an imposing courtyard, a jewel in the crown of equestrian architecture. This unique, two-storey octagonal building was so perfectly planned to fit the 18th century courtyard that it radiates a sense of satisfaction and well-being.

Milton Hall in Cambridgeshire is home to the Naylor-Leyland family whose Fitzwilliam antecedents bought the estate in 1502. Sir Philip is currently Master of the Fitzwilliam (Milton) Hunt, one of the great ancestral foxhunting packs in England.

The stable courtyard was built in 1690 by William Talman and extended in 1720. The two-storey buildings provide for carriages, some twenty horses, hay lofts and staff accommodation and are built of a pale limestone, probably from the local quarries of Ketton or Marholm. The yard itself is of granite setts. The building's octagonal design in the centre of the courtyard is as practical as it is handsome, allowing easier movement of horses and carriages around its facets, so avoiding the hazardous corners of a square or rectangle.

This historic place has seen many changes. The stable yard clock tower was recently repaired and several bullet holes were found in the weather vane. These were from a pistol shooting range in the nearby kitchen garden used during World War II by the Czech army who were stationed in the stable courtyard. The Special Operations Executive (SOE) was also here in 1944 before D-Day preparing for Operation Jedburgh, a plan to co-ordinate the resistance network in Europe prior to the allied invasion. The author Daphne Du Maurier used to visit Milton Hall as a child and is said to have based her description of the interior of Manderley in her novel *Rebecca* on Milton Hall.

I visited the saddle room on a bright April day after the hunting season had finished so all was quiet. I opened the heavy door and sunshine flooded in. I was unprepared for quite such an exceptional and beautiful room; it was not just a service area for a busy yard but more of a monument intended to venerate the past and to celebrate the Fitzwilliam Hunt. It was built with confidence in the future.

It has an innovative heating system which originally extended out to the surrounding stable wings. The coal-fired boiler in the cellar still heats a huge central pipe which, like a ship's mast, runs up through both floors of the tack room, the top floor

The octagonal saddle room at Milton Hall set in the 17th century courtyard and built of local Cambridgeshire limestone. Architectural detail of the period such as the 'oeil de boeuf' oval windows can be seen in the staff accommodation wing behind. The original carriage bays are to the left and stabling along the south and west wings completes the quadrangle.

Stirrups for the hunt saddles, all clean and ready to go – just as they have been for at least two hundred and fifty years.

being the main drying room for rugs and hunting equipment.

The ground floor houses the hunt saddlery. Light streams in through large Georgian sash windows and gleams on the red and black floor tiles. Around the wood-panelled walls are forty spotless saddles, stripped of girths and stirrup leathers which are hung separately below.

These are traditional English leather hunting saddles of various makes. The bridles are of typical hunting-weight leather with curb or snaffle bits, cavesson and drop nosebands and running martingales.

On one broad window sill are all the stirrups, on another there is a pageant of old hunting equipment: a collection of leather cases to attach hunting horns and wire cutters to the saddle. Bowler hats, old receipts on a wire hook, a map of the hunting country and photographs are reminders of the long history of this distinguished hunt.

A sun-yellowed copy of the poignant 'The Prayer of a Horse' by Milton Bode is pinned to the wall. It is printed 'With the compliments of Day, Son & Hewitt Ltd.', a manufacturer of animal medicines (including 'the Black Drink' an equine tonic in 1833) awarded the Royal Warrant by both Queen Victoria and Queen Elizabeth II.

Among the fittings is an elegant saddle horse with panelled storage cupboards below, made of a polished hardwood, on cast-iron wheels. It was probably custom-built to support driving saddles for the carriage horses while they were cleaned; the unusual concave top would fit their padding better than it does a riding saddle. No maker's name is evident but movable saddle-horses of this fine quality are rare.

A smart array of saddlery at Milton Hall. A long-handled, two pronged steel fork is used to put up and take down the saddles.

Musgrave of Belfast produced portable saddle horses around the 1900s of excellent design. These are sought after by collectors, particularly from Europe, and are seldom seen in British tack rooms now. They have a cast-iron frame on wheels and a pitch pine body. The A-shaped top which supports the saddles opens out to form a flat working surface – the stable staff probably had their lunch and played cards on this while the horses were out before quickly folding up the sides to receive the saddles as soon as they returned. On some models the top was detachable and when reversed supported the saddle upside down so that its underside could be cleaned. There are drawers at one end and a cupboard for cleaning materials at the other.

This saddle room is likely to have seen the transition from stag and hare hunting to fox hunting and has remained unaltered, keeping an unbroken continuity from the early 18th century to the present day. In the tradition of British hunts, the Fitzwilliam activities are interwoven with the local community and the busy hunt calendar continues year round with point-to-point racing, hound and horse shows, conservation work, fallen stock disposal services for farmers and fund-raising for a wide range of charities.

The Fitzwilliam (Milton) Foxhounds

The management and breeding of hounds is both an art and a science and central to the tradition of hunting. The foxhound pedigrees here, detailed for every hound, go back to 1751 and an even earlier book of records was lost in a fire.

A pack of hounds that could provide sport all day was highly prized. Individual hounds were considered to be important enough to be painted by the finest artists of the day, such as 'Brocklesby Ringwood 1788' by the celebrated George Stubbs (1724-1806) and 'Bluecap' for the Cheshire Hunt by Francis Sartorius' (1734-1804). Bluecap was an exceptional working hound and had a national reputation; he was timed during a speed match in Newmarket to have run four miles at an average speed of thirty miles an hour. He sired many litters for the Fitzwilliam, Belvoir, and Brocklesby Hunts and died aged thirteen in 1771. An engraved obelisk was erected in his memory and the local pub in Sandiway, Cheshire, changed its name to The Bluecap.

The prestigious Peterborough Royal Foxhound Show is now in its 130th year and the Earl Fitzwilliam was its first president in 1878. The high standard of entrants at the show reflects the meticulous breeding of these healthy working hounds, free from the hereditary problems commonly found in non-working show dogs.

Hounds are judged on criteria relevant to their work: on good conformation for pace and stamina, for example, good feet, straight legs and deep chest for a sturdy heart and lungs. Dog hounds and bitches for breeding are also selected on the quality of their scenting ability, speed and voice in the hunting field. Stud hounds are drafted between packs to confer particular characteristics and no money changes hands. Types of hounds vary with the terrain, for example larger, broken-coated and more independent hounds, known for their good voices, hunt in the Welsh mountains while the majority (Old English and Modern English) in the Shires have smooth coats of black and tan on white. It is thought that foxhounds are descended from the original Norman hounds which relied on sight to pursue stags and hares. Gradually other breeds with scenting ability were bred into these lines to produce the types of foxhound we see today.

About a mile away in Milton's park, the hounds are kennelled in an extraordinary building which was built to resemble a mock ruined castle in the 1760s. Nearby are ancient oak trees from this period (and possibly as early as Tudor times) which still survive and with longhorn cattle grazing beneath them it is easy to be transported back to the tranquillity of the 18th century. James I is recorded as hunting deer over this countryside in 1615 from nearby Apethorpe Hall.

Today, this heritage scenery contrasts sharply with the vast housing estates recently built around nearby Peterborough which, together with new road networks, have reduced the hunting country to about six hundred square miles. Before 2004 when hunting with a full pack of hounds was still legal in England and Wales, hunting two or three days a week meant, at a rough estimate, that each area of this country was visited by hounds on five days in one season, leaving 360 days of that year for the wildlife to live and breed around the farmland. This rate aimed to achieve a balanced fox population by culling old and weak animals and to keep the numbers of foxes in proportion to their food resource.

The kennels building has recently been extended to provide excellent accommodation for about fifty couples of hounds. Weaned hound puppies or whelps are lent to hunt supporters so that the young hound becomes socialised and learns its name and basic manners. This is known as 'walking' a hound puppy. After about eight months the hounds return to live as a natural pack at the hunt kennels. Many hounds and their 'walkers' bond for years afterwards.

Fitzwilliam Hunt hounds at the Milton Kennels. Dogs and bitches live as a pack and hunt together, except when breeding. Each knows its name and is selected according to its abilities in the terrain for a particular day's hunting. For example, if large numbers of deer are anticipated in that part of the country, older hounds who are less likely to be confused by different scents may be preferred. Those with a good voice would be more useful in wooded country.

Young hounds are 'entered' or introduced to the hunt in the autumn and learn from the experienced older hounds.

Every hunt has an annual Puppy Show which is not just a social occasion to thank the puppy walkers for their work but keeps a check on the quality of its pack and provides a guide to future breeding lines.

Upstairs at the Milton kennels, in one of the turrets is a cosy octagonal room (reflecting the shape of the saddle room) with a couple of easy chairs and murals of prizewinning hounds of the past. The huntsmen's red coats hang clean and ready, each with six silver buttons at the back bearing the Fitzwilliam family crest. Beside them on a peg are the special stirrup leathers worn across the chest. This was started by the Earl Fitzwilliam who broke a stirrup leather during the chase, and requested that his hunt staff should always carry spares.

The Fox

Central to the hunting debate is the fox and its welfare and the need to maintain a population balance of healthy foxes in the countryside. The European red fox (*Vulpes vulpes*) is at the top of the wildlife food chain and lambs, particularly in the upland regions of the UK, are vulnerable to its predation. Despite its notoriety as a wanton killer of chickens, the sight of a fox crossing open country can be a rare treat and gives a frisson of wonder and respect. This handsome and intelligent wild creature is one of very few with enough confidence to 'cock a snook' at human civilization. There are witnesses to its bravery too; in a narrow Devon lane the Tiverton hound pack met a fox coming towards them which jumped up, ran across their backs and escaped. During a chase in Cheshire a fox calmly swam across the Shropshire Union canal while hounds went roaring past on the towpath. Another time, a fox ran through the middle of Chester Zoo causing chaos among the following hounds confused by an explosion of exotic scents. Foxes have been woven into our folklore and storytelling culture from Aesop and Chaucer to Nash's ballad 'The Fox's Prophesy' and books by Beatrix Potter, Roald Dahl and Douglas Adams.

HUNTING

The fox's habitat has significantly altered since World War II as the countryside has been squeezed by urban spread. Other factors include an increase in the badger population (since they received protected status in 1992) which may affect the fox population as both species compete for similar food resources.

Changes in farming practice, such as the autumn sowing of winter wheat and the widespread spraying of herbicides, neither of which favour wildlife, have also caused the opportunistic fox to migrate to the towns. Here there are easy pickings from fast food chains and household bins and the teeth of urban foxes, often rotten from this unnatural diet can predispose them to sarcoptic mange which causes great suffering. Urban foxes which are 're-homed' to the countryside do not necessarily retain the hunting skills to survive there.

The release of zoo animals into the wild, even the return of zoo-raised Przewalski horses onto the steppe, has shown that animals who are accustomed to having their food brought to them, soon lose their wild survival skills. The stringent quarantine veterinary regulations for dogs in the UK have kept the fox population here free of rabies since 1922 though it is still found in Eastern European foxes.

The selective culling of foxes by hunting with hounds helps to keep the fox gene pool healthy. The hunt dispatches the less fit foxes providing a quick end to what might otherwise have been a lingering death from starvation due to gunshot wounds, poisoning, mange or old age. A hound, which is twice the weight and size of a fox, kills it swiftly by breaking the neck, as would a dog catching a rabbit. The welfare of the fox in this respect has been studied by the Veterinary Association for Wildlife Management (2007) which is in agreement with this view, as is a special report by the Game Conservancy Trust (2000).

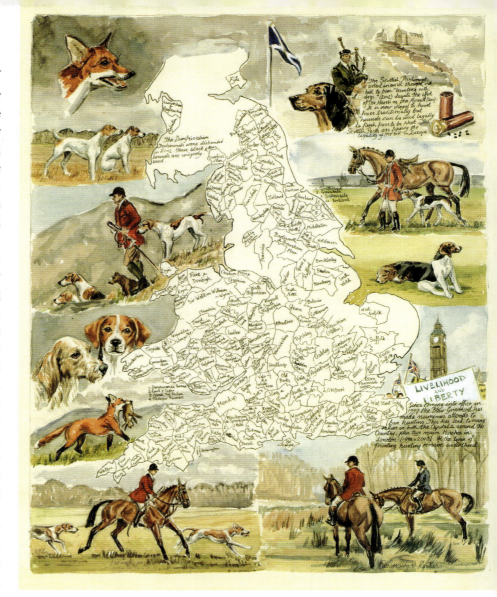

Map of hunting countries of England, Wales and Scotland in 2004

The Aims of the Hunt

Before 2005, the aim of hunting was to keep the livestock farmers happy without reducing the fox population beyond recovery so maintaining the intricate network of the countryside in balance.

There was no hunting during the summer. Traditionally this was to allow for the growing and harvesting of crops on arable land and to respect the breeding season of the fox. The foxhunting season resumed in the autumn when the cubs had grown to maturity and become familiar with their home territory.

Since the ban, an important part of the hunt's work continues to be in conservation during the closed season in late spring and summer. Copses are planted and managed in order to encourage all forms of wildlife and to provide sufficient prey and breeding areas for all levels of animals in the food chain, including rabbits and foxes.

Good farming practices were always encouraged, such as leaving a wide margin or 'headland' around cultivated fields which allowed small hedgerow mammals to forage close to the safety of the hedge. This headland, originally created by leaving turning space for the team of plough horses at the end of the furrow, is important for wildlife and in hunting country is retained. In non-hunting country, mechanised farming is so efficient that not only are the headlands eliminated but very often the hedges too. Any damage done by the hunt is repaired by its fencing team. This is one of many jobs, such as those for farriers, saddlers and vets, supported by hunting.

The riders hunt for many reasons. Hunting provides access to and a deep immersion in the beauty of our landscape. The frosty breath of horses in the winter sunshine, the scent of autumn leaves and the patterns of farming laid out like a painting can fill the soul brim full.

Many go to experience the thrills and exhilaration unique to riding at speed over hedges, walls and ditches on an unpredictable line across the countryside. Others go for social reasons to enjoy riding off-road with friends, to accompany a child new to hunting on a leading rein or to observe the scenting skills of the hounds as they work. Riders who prefer not to jump can follow by finding a way round through gateways and many enthusiasts follow the hunt in cars and on foot. For the horses, running with the herd is natural and produces a level of excitement not generally shown when out hacking.

Before the ban, an alpha fox returning from visiting vixens outside his territory could lead hounds for several miles before outmanoeuvring them. After such a run late in the day the master sometimes decided to 'give it best', and ask the huntsman to call off hounds from the pursuit. Different calls on the huntsman's horn are used to convey instructions to the hounds and also to inform everybody present. The call of a copper horn about nine inches long can carry across a couple of miles, a silver horn slightly less far while the 'speaking' of hounds carries the furthest for about three miles on the breeze.

Today, headcams worn by riders can provide a video recording of all the day's work but records used to be kept differently.

From descriptions heard from hunting people decades ago, after the day's hunting was finished and the horse rubbed down and fed, the master would sit down at his kitchen table with a glass of whisky, a large map and a piece of string for measuring. In his journal he would record the name of every covert drawn and the distance of the 'point'. This is the mileage during a chase in a direct line from start to finish, as the 'crow flies'. Of course, the actual ridden distance would be far more than this and, if hacking to and from the meet as well, could be twenty miles or more on a good day.

One of the most famous points was the 'Greatwood run' by the Duke of Beaufort's hounds in 1871. They achieved a point of sixteen miles and covered a total of twenty miles in three and a half hours. The fox crossed the River Thames twice before safely going to ground. Even this though did not come close to the unrepeatable 'Grand Chase' by the Duke of Richmond's Charlton Hunt in 1739 which measured fifty-seven miles and finished up in the dark – no doubt with the use of many changes of horses by the survivors.

Hunting also provided a unique sporting challenge which required skill and courage. The Great Waterloo Run with the Pytchley in 1866 illustrates the extraordinary calibre of their master, Colonel Jack Anstruther Thomson. A fox was found at 2pm and followed for twenty-six miles. The Colonel, a large man weighing sixteen stone, outlasted both his whippers-in at the finish, despite four falls and changing horses five times. By then it was 5.30pm on a dark February day and the master hacked back nineteen miles to kennels and arrived at 10pm, handed over his horse and went on to the hunt ball at Market Harborough.

Before the ban, it was unusual for the mounted followers, known as 'the field' to witness the killing of a fox by hounds. Every rider has to keep behind the 'Field Master' at all times so as not to interfere with hounds working. If this rule is ignored, or if a horse kicks a hound, the rider may be sent home. Trail hunting today still provides an education for young horses which may go on to have other careers such as racing or eventing, as well as a second chance for older horses which might otherwise be surplus.

The regulatory governing and disciplinary body for hunting is the Masters of Foxhounds Association which represents about 174 hunts. Generally, each hunt has a Master (or two or three Joint Masters) who liaise with the farmers and plan the day's hunting, a huntsman who is in charge of the hounds and is aided by a whipper-in or two and a field master (either the Master himself or a senior hunt member appointed by him or her) who leads the mounted followers across country. Kennel and stable staff are also employed to maintain the hounds and hunt horses. The costs are borne by subscriptions from members, private donation and funds raised by the Hunt Supporters committees.

Two other organisations are important: the Hunt Staff Benefit Society, established in the 19th century as a benevolent fund to provide for retired hunt staff, and the various supporting Hunt Clubs. The oldest of these is the Tarporley Hunt Club in Cheshire, founded in 1762, who still meet for dinner in the magnificent panelled Hunt Room at The Swan, a 16th century coaching inn, wearing the traditional scarlet coats with green collars.

In the last 300 years hunting has weathered many setbacks such as changes in the law, foot and mouth epidemics and two world wars. It continues to be a valued and vigorous part of rural life and to draw together supporters young and old from many different backgrounds.

The Fitzwilliam Hunt testimonial meet at Milton Hall to celebrate George Adams' 34 years of service to hunting on the 2nd March, 2016. Left to right: Josh Worthington Hayes (2nd whipper-in), George Adams (huntsman) and Simon Hunter (1st whipper-in).

CHAPTER TWENTY-ONE

Horse Logging

The use of horse power in forestry is as old as house building, mining and ship building and was the only way timber could be extracted for use as beams, pit props, keels and masts until the 1930s. The arrival of the combustion engine in the form of the tractor after the Second World War put most heavy horses out of business in agriculture and forestry for the next sixty years.

Very few remain in agriculture today but a number are being used again to help harvest timber. The horse's light hoof print, in contrast to the heavy tread of powered machinery, is now particularly valued for the sympathetic management of our woods and landscape. A horse has the advantage of working selectively to extract individual trees whereas a machine is best at taking out lines or racks of trees.

In the 1990s, timing and talent came together with the increased public awareness of the environment and a pioneering group of people who formed the British Horse Loggers. Foresters learnt how to handle horses and horse handlers learnt forestry skills, each helping the other and they began as a specialist group within the Forestry Contracting Service. They followed the European Draught Horse Federation's guidelines on the care and use of traction animals as modern horse logging was already well established in parts of Europe.

Two of the British Horse Loggers founders, Richard and Angela Gifford, who were experts in the teaching and practice of working horses, introduced new techniques to the forestry industry in Britain. They travelled to Scandinavia and returned with a horse-drawn arch (*see* page 144) for extracting logs, engineered by Per Arne Ulvin in Tangen, Norway. In 1995 they exhibited it at the Royal Agricultural Society of England Show and were awarded the Duke of Edinburgh's perpetual trophy for horse-drawn machinery and the Show Society's silver medal for design.

The arrival of the arch was a leap forward for British horse logging and was a key piece of equipment in the forestry industry's recognition of horse logging as a viable commercial method of extraction. The use of Ulvin's arch proved to be safer and more efficient than using chains (long gears) as explained below. Since this prototype, new arches have been designed and Morgan Andersson in Sweden offers a variety of horse-drawn machinery such as single horse forwarders and two horse forwarders with a mechanical grab for loading and unloading logs.

Under the inspirational chairmanship of Doug Joiner (2000–2013), the British Horse Loggers (BHL) became an independent organisation with HRH the Prince of Wales as its patron, the Prince having an active interest in working horses (including his own Suffolk Punches) on his Highgrove Estate.

Horses of many different draught breeds are used, in fact, ten different breeds were counted at the 2017 annual British Horse Loggers competition. Dales and cobs are popular and continental breeds such as the Ardennes, Comptois and Belgians are known for their good temperaments and stamina. More important than the breed is the horse's temperament; a quiet and steady nature is essential. The horse must be comfortable with the noise of

A team of loggers and their six horses in Denbighshire, North Wales. Date unknown.

chainsaws and falling trees, be able to stand reliably even when unattended and be willing and able to do the work.

The object of modern horse logging is to move logs from the felling site with the least amount of environmental damage, to the landing where they can be collected by lorry. A horse leaves little trace of its work and is therefore a good choice for working on ecologically sensitive sites with fragile flora and fauna, such as orchids and dormice in ancient woodland. In fact, the scarification of the woodland floor caused by a working horse actively encourages natural regeneration whereas heavy machinery can leave deep ruts and causes greater disturbance with noise. Engine-powered forestry machinery may also pollute the ground with fuel, compact the soil and alter the drainage. It may be limited by steep gradients and requires good access but it is also efficient, especially in clear felling situations, usually having a higher productivity than a horse. The capital outlay for such machinery is enormous compared with a horse logger's requirements, though often horses and machinery can work well together, each having their particular advantages, depending on the site.

Two common methods of extracting logs with a horse are a) in shafts with an arch or a wheeled vehicle (forwarder) and b) in long gears with chains and a swingle tree (*see* Harness page 34).

The design of the Ulvin arch (Kombidrag) has been much copied and modified. It has two independent shafts, each with a wheel and skid at the rear and a slot to attach to the horse's harness in the forward position. Instead of a straight axle connecting the two wheels, there is an arched steel bar which carries three or four winches. It can be dismantled for easy transportation and quickly assembled on site.

A short chain is secured round the butt end of the log, and passed under the cogs and onto the manual winch which raises it off the ground. Once the logs are secured on the winch, the operator has complete control of the load. A variety of timber diameters can be chokered and extracted in each load with eighteen inches being the limit for the height of the arch.

When the horse moves forward in the shafts, only the tips of the logs are in contact with the ground. This reduces the friction and load for the horse and minimises damage to the forest floor.

Harness designed for use with the Ulvin arch is made in Sweden by Tärnsjö of vegetable tanned leather (*see* page 1) with stainless steel fittings. The collar is fitted with adjustable wooden hames and fastens under the horse's neck. The shafts, tug straps and breeching connect to a traditional Scandinavian designed junction ring on the harness.

It is, however, very expensive and American nylon webbing harness is often preferred because it is strong, cheaper and lighter than biothane or leather and easy to clean.

Horse loggers are a group of individualists: their horses, harness and their work site conditions vary widely and each logger adapts and improvises accordingly. The Blue Horse Equine Ltd. company has provided British loggers with access to a wide range of American harness and horse-drawn machinery and also trials new equipment such as synthetic ploughing collars. Since horse work is so popular in America (a Horse Progress Event in the US in 2017 attracted 30,000 visitors over two days), it is a good source of innovation from which we and our horses can benefit.

Trace harness (long gears) consists of chains attached to the horse's collar which run backwards through loops on the harness and are connected to a swingle tree to which the log (or load) is attached. The log is in contact with the ground throughout its length and needs to be steered round stumps or boulders. However, since horses are frequently asked to work in restricted spaces which are too difficult for operating machinery, long gears are much more appropriate and efficient than an arch.

For working in long gears with chains (traces) and a swingle tree, plough harness with collar, hames and meter strap, belly band and crupper can be used. Springs may also be incorporated into the chains at either end of the swingle tree to absorb any shock when the horse starts off with a heavy load.

The bridle is often worn over a headcollar (useful for tying up the horse during breaks) and the bit is chosen to suit the horse. The use of blinkers is optional. Once a horse has been trained, most horse loggers prefer to use an open bridle allowing all-round vision for the horse. Blinkers may inadvertently guide a branch into the eye rather than protecting it.

Until his recent retirement, Jim Johnstone was a well-established professional horse logger working in Dumfriesshire, Scotland. Jim also taught at Barony College for eight years and was an instructor on horse logging, machine and chainsaw operation.

As Jim pointed out, horse logging is not romantic, it is tough work and potentially hazardous. Much of the work is done in the autumn to early spring period; outside the bird nesting season, when there is little foliage and the moisture content of timber is low. Winter weather can be harsh and horse flies, ticks and midges are summer irritants.

A typical contract for Jim would be from a commercial timber management company, such as United Paper Mills Tilhill, in which five hundred tons of larch thinnings had to be extracted from a steep and ecologically sensitive site where heavy machinery could not be used. Jim's 17hh Belgian draught horse, Billy, removed eighty tons of this timber. The rest was carefully extracted using mechanised forwarders which lifted the logs clear of the ground and stacked them where they awaited collection by

Four horses in trace harness moving logs near Nantglyn in Wales. Date unknown.

The late Doug Joiner, Chairman of the British Horse Loggers (see page 140, 148) at The Stiperstones in Shropshire. His Percheron mare, Ella, is working in long gears and an American collar. Doug is using a hawthorn tree to scarify the ground and encourage heather to grow for the 'Back to Purple' environmental project.

lorry. Billy was brilliant; when he was only six years old he could work on his own taking thirty-foot-long poles down a slope to the landing but would stand like a rock when required. 'He's the perfect horse', Jim said, 'He has correct conformation, strong and compact with twelve inches of bone and weighing about 950 kilograms. He can pull out ten to twelve tons of cross-cut logs in a day without breaking into a sweat.' For this type of work Billy would eat twenty-five pounds of oats and barley mixed with alfalfa every day. A midday break of an hour and a half in the woods is routine: time for a horse to drink, feed and browse while the logger gets out his Thermos and sandwiches, and makes sure that a hot horse doesn't chill while standing.

Steffi Schaffler (BHL committee member) and David Roycroft are full-time horse loggers and foresters in south-west Scotland. They have a Comptois, an Ardennes and two Clydesdale cross breeds and prefer to train their horses when young. By having a range of skills they can work all the year round: felling, extracting logs, firewood processing in the winter and bracken-bashing and mowing in the summer. Rolling contracts from year to year on local estates and farms and their own rented fifteen hectares of mixed woodland from the Scottish Woodlot Association provides good work – enough to consider taking on apprentices in the future.

The surrounding countryside is steep and wooded and they use a variety of equipment: an eight-wheeled Morgan Anderssen forwarder, an Ulvin arch and long gears with a swingle tree. Because of the diverse terrain they may need to use different equipment throughout the day and find a nylon American 'D Ring' harness the most adaptable and easiest to maintain. The 'D' ring is shaped like a shackle and is more versatile than the round junction ring on the Ulvin arch harness. This American

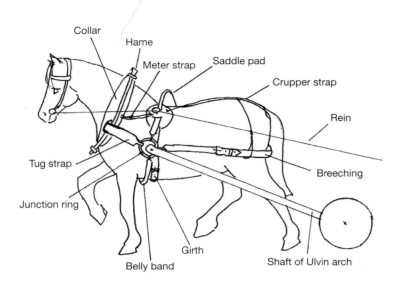

Harness for Ulvin's logging arch

harness is suitable for trace work with a swingle tree, has shaft loops for English shafts, can be used for pairs work with a pole and also accommodates the Swedish shafts on the Ulvin arch and forwarder.

Building a reliable partnership with a horse takes time. Most horse loggers would identify with the satisfaction of being in such harmony with their horse that instructions are almost superfluous. Every horse logger has a wealth of stories about their equine partners. Here is just one personal example.

I wanted to learn an additional skill with a horse which was beyond riding. A talk at an agricultural meeting about forestry work with horses started me off. I had no previous experience of draught horse work but when a window of opportunity arose in a busy career, I jumped at the chance to learn.

Luck was with me in the choice of my tutors: Richard and Angela Gifford, Marcus Van Stone and Kirsty Robb. The Giffords were experts in the draught horse world and pioneers in British horse logging. Marcus and Kirsty were consistent winners of the Horse Logging Championships in England and Scotland. For a brief period, all four taught horse logging courses at an agricultural college fifty miles away from where I lived. They were expert in explaining theory and practice and gave me the confidence to set up on my own. I planned to extract 'thinnings' with a horse and leave the felling and advanced logging to those who were qualified. My mother agreed to look after a horse, if we could find one, on her smallholding in Wales.

After the courses I spent a week shadowing Marcus who was working on an estate in Inverary on the west coast of Scotland. The trees, mostly Sitka spruce, were huge, the rain was torrential and Marcus' skill with his horse, Campbell, was impressive. Some months later, I bought a twelve year old Shire-cross mare, Dolly, from two estate workers in Wales – she was 16.1 hands high, black with a white blaze and two white stockings. She had been bred by gipsies near Manchester, sold for riding and had done a little work moving logs on the estate with long gears. Dolly communicated her opinions as clearly as though she were speaking and although exuberant at times, was affectionate and sensible.

I would take her up on the mountain, walking behind her with long reins for miles while we got to know each other and became fit enough for work. Dolly and I learnt together by trial, error and improvisation with guidance from the British Horse Loggers group, who at that time were still associated with the Forestry Contractors Association. I heard Hans Sidback, Sweden's premier horseman, talking at a BHL meeting: 'Working with a horse is about empathy, mutual respect, teamwork and trust and a calm, consistent and confident handler', he said and I aspired to those principles.

We graduated from pulling logs with long gears to using an Ulvin arch, Dolly learning to do a neat turn on the haunches in the shafts by crossing her forelegs. The local saddler helped me to improvise a working harness for the arch with an English ploughing collar, two gate hinges modified to attach the shafts and parachute harness clips to attach the breeching. This very unorthodox arrangement worked well until we earned enough to buy the Swedish harness designed specifically for the arch.

Dolly was patient with my mistakes, and looked after me for the next ten years, both of us remaining accident-free. I learnt to read her body language, she learnt what was expected of her and we trusted each other. We shared a vocabulary of about six words: instructions which she would usually follow even if I was some distance away from her.

When using a swingle tree (steel bar on right), it is advisable to have a 'safety link' in the harness to quickly separate the horse from the load (connected to chain on left) if either develops a problem. This is an old forged iron design, where the nose of the link to the swingle tree is held closed in a groove with a movable iron ring (left). The ring can be slipped free to open the link (right), so that the load may be disconnected even if it is under tension. A cord can be threaded through the eye on the hook and the eye on the ring to hold the ring in position during normal work. Alternatively, the hame strap on the collar can be cut or undone so that the horse can be relieved of the load.

Every job was different: on private estates, in council-owned woods, Forestry Commission projects, National Trust conservation, demonstrations and shows, each was a new experience. Every wood had its own character; some were broadleaf with beech and oak and wild garlic in the spring, others were coniferous, densely planted on ridge and furrow ground. These were dark and difficult to manoeuvre in and the brash was dagger-sharp but the thick feathers on horses' legs protect them from harm. Some woods had deeply muddy access tracks, others grassy rides.

Our first job, only three weeks after Dolly had arrived, was to pull out some large, freshly felled oak logs for the artist in residence in Hartford, Cheshire. The logs lay on a steep slope below a playing field which was crossed by a busy railway and surrounded by council houses. The task was to extract the logs using the Ulvin arch, take them a quarter of a mile to the end of the field and stack them there. We finished in the dark and my helper left for home. As I was unharnessing Dolly by the trailer, we were spotted by a crowd of children aged about six. They swarmed out of a nearby house dressed in party dresses, white socks and ballet slippers. Then the boys descended on my logging chains and ran off with them while the little girls entwined themselves round Dolly's legs and played with her tail. No adults came to my rescue. Fortunately she stood like a rock, the gipsies having done a good job with her upbringing. Eventually I was able to disentangle the children so that Dolly could move without treading on them and we loaded up by torchlight and drove sixty miles home. Our partnership began that day. I had to return the next morning to find my chains which were buried in the mud.

At work in the woods, the heavier the load of logs (up to half a ton) the faster she would go to keep up the momentum. Her fast trot was as fast as I could run behind her while holding the long reins. If I tripped on stumps or caught a foot in brambles I fell headlong hanging onto the reins and was in danger of being dragged. But she always stopped and waited until I was upright again and she remembered the geography of each wood better than I did. We would arrive at a new job and walk the route together twice, noting where the felled logs were, where there was room for turning and where the extraction route led to the landing. Sometimes, in a dense wood and walking behind Dolly to pick up the next load, my view of the way ahead was blocked by her large black hindquarters. I would indicate where I thought we should turn to follow the path, but more than once she was keen to continue to the next turn where I realised that she was right and I had been wrong. We developed a rhythm: a quiet phase when Dolly stood still while I chained the logs and winched them up, then an intense physical effort dragging the load to the landing site, followed by a breather for her while standing again as I released and stacked the logs.

For some jobs I had to leave her overnight in the wood in an electric fence enclosure while I returned home. Early the next morning she would greet me with a whinny looking for her breakfast. It was agreed between us that the daily shift would end at 4pm or earlier in the short winter days. She would stop and look pointedly in the direction of our base at about 3.45pm, her natural clock always keeping perfect time.

We were asked to demonstrate to the primary school children of Snowdonia how to take out thinnings as part of an imaginative three day project for National Book Week. The children would be shown how to plant an acorn, how to select trees for thinning, observe the felling and Dolly and I extracting them. They then watched how the pulp was processed and were told how it was made into paper for a book.

The wood was close to the Penrhyn slate quarry which blasted twice a day at that time. During my site visit, two F16 fighter jets roared out of the Ogwen Valley just above the tree canopy with no warning, training for the Iraq war. This would be a challenge for any horse but after some preparation the risks were reduced to an acceptable level. We agreed with the quarry supervisor to work round the blasting times and my request to the RAF to divert training flights for three days was willingly granted. Maggie Braunton, an expert horse logger and Ruaidhri, her Ardennes, joined us on the second day and the project was a great success.

Being part of the restoration plan for neglected woodlands in suburbia was rewarding. These areas were usually full of litter, with watercourses blocked by old shopping trolleys and trees that had been damaged by carvings in the bark and set on fire.

Jim Johnstone's (see page 142) horse logging tack stored in his barn. He prefers to use Canadian Amish harness with collar pads. Synthetic harness (biothane) is more practical than leather in this high rainfall region as it dries out quickly, weighs less than leather and needs little maintenance.

Billy, a 17hh Belgian draught horse, working in Auchencairn Forest, Dumfries with Jim Johnstone. The Canadian Amish collar and biothane harness (long gears with swingle tree) has quarter straps with the breeching for shaft work.

Programmes in which the community was invited to participate in the woodland restoration were successful, especially with the children. The presence of a working horse was popular and provided good publicity for the local council's efforts to create a valued amenity. During these 'Country Days', perimeter hedges were cut and laid, new hawthorn slips planted in the gaps and demonstrations of pole lathing, gipsy rose whittling and stick carving were arranged. Bird and dormouse nest boxes were put up and fungus forays, coppicing and bird watching were part of a continuity programme run by rangers. On these occasions I was often asked some revealing questions about the horse by the local children, for example, 'Please Miss, does it run on petrol or diesel?' Another time, when the horse was steaming gently after working hard on a cold day, a concerned child grabbed my arm and shouted, 'It's on fire, Miss, it's on fire!'

Dolly died aged twenty eight after a happy working life. I can say that because it seemed to be her opinion. She lived out with other horses in the field but in the winter came in for a morning feed. If we had a logging job to go to, the trailer would be drawn up in the yard ready for loading with the ramp down, on the right hand side of her loosebox. When her door was opened she had a choice: to turn left and rejoin her companions in the field or to turn right and go to work. She always turned right.

One of her companions was Neville, a Clydesdale cross, who came off the boat from Ireland with no papers or history, went to a riding school and then to a family before he found us. Opposite in temperament to Dolly, being quite private and uncommunicative, he eventually became an excellent logging horse.

In his retirement he thoroughly enjoyed being long-reined on walks, carrying a picnic in panniers for

Machinery and horsepower working in harmony in south-west Scotland. The horse is harnessed to an Ulvin logging arch which has just delivered logs to the landing. (Photo courtesy of Steffi Schaffler)

a group of friends and grazing unrestrained nearby while we had lunch. Both he and Dolly gave years of pleasure, not just earning their keep by working but being ridden, pulling a sledge for children in the winter and doing odd jobs on the smallholding.

The British Horse Loggers is a thriving national organisation with approximately seventy members and a professional register of approved contractors (full and part time). A separate charitable trust, career apprenticeships and award schemes have been established.

It offers practical support in many forms both to members already working horses in woodland environments and to those wishing to begin. Courses on all aspects of horse logging are organised around the country as well as educational exchanges with horse loggers in Europe. Horse logging has become a recognised and respected branch of forestry and environmental management and a type of work very well suited to draught horses.

Fly Repellent for working in the woods

2 tbsp. Vinegar
1 tbsp. Baby Oil
1 tbsp. Citronella Oil
1 tbsp. Neem Oil
1 litre Lemon Tea
Mix together and shake well.

Logs to Burn

This anonymous old English poem may be read with a pinch of salt; wood for burning should be seasoned for two years if possible and resinous wood which spits is best enclosed in a woodburning stove. According to the poem, elm is not valued in Britain as firewood yet elm (and birch) are highly prized in Norway. However, since Dutch Elm disease arrived in Britain in the early 1970s, it is not easily available here.

> Beechwood fires are bright and clear
> If the logs are kept a year,
> Chestnut's only good they say,
> If for logs 'tis laid away.
> But ash new or ash old,
> Is fit for a queen with crown of gold.
>
> Birch and fir logs burn too fast
> Blaze up bright and do not last,
> It is by the Irish said
> Hawthorn bakes the sweetest bread.
> Elm wood burns like churchyard mould,
> E'en the very flames are cold
> But ash green or ash brown
> Is fit for a queen with golden crown.
>
> Poplar gives a bitter smoke,
> Fills your eyes and makes you choke,
> Apple wood will scent your room
> Pear wood smells like flowers in bloom
> Oaken logs, if dry and old
> Keep away the winter's cold
> But ash wet or ash dry
> A king shall warm his slippers by.

Anon.

Glossary of Logging Terms

Brash: Twigs and small branches left on the forest floor after the main branches have been lopped off or 'snedded' from the main trunk.

Thinnings: These are young trees which are felled and removed to allow better quality trees to thrive. Second thinnings may be taken out at a later time, depending on the species of tree and the rotation time (planting to felling) of the crop.

Lines: Also called long reins and made of rope or leather about twelve to fourteen feet long. The lines are connected to the horse's bit, threaded through rings (terrets) on the harness and held by the handler who guides the horse from behind.

Long gears: This is a simple harness used by a horse to pull a log or bundle of saplings along the ground, also called snigging or tushing. Chains are attached to the horse's collar, and extend backwards on each side through supportive loops on the harness to a swingle tree to which the load is hooked on.

Landing: An area where the extracted logs are stacked awaiting collection by lorry, usually alongside a road, hard track or on a hard surfaced area (hard standing).

Rides: A ride is a permanent hard or grassy access track through to the stands of trees or wooded areas.

Rack: A route for extraction formed by cutting out a line of trees in a planted stand of trees.

CHAPTER TWENTY-TWO

Long Riders

'Long riding is not just travel, and horses are not just a mode of transport, it is a way of living, a relationship with the earth, with people and the greater web of life that inevitably rewires horse and man.'

This was written by the Australian, Tim Cope, who rode unaccompanied from Mongolia to Hungary. His was the first modern crossing of the Eurasian steppes on horseback, beginning in 2004 and taking three years. This was an epic journey of over 6000 miles.

The 13th century Mongol horsemen rode faster and further than anyone else. Their superior horsemanship and the efficiency of their relay stations, where tired ponies could be swiftly exchanged for fresh ones every forty kilometres, were ideal for communicating military intelligence. Ghengis Khan's general, Subotai, is said to have travelled 2000 kilometres from the Caspian Sea to Samarkand in a week.

This relay system was the forerunner of the legendary Pony Express which carried the news and mail from Missouri to California until the introduction of the telegraph.

An 1860 advertisement for riders in St Joseph, Missouri reads, 'Wanted. Young, skinny, wiry fellows not over eighteen. Must be expert riders, willing to risk death daily. Orphans preferred.' Successful applicants were equipped with a bible, a pistol and the leather mail pouch and each rode some 80 miles changing ponies seven or eight times. The journey of about 1800 miles to Sacramento, California, was usually completed in ten days, weather and Indian attacks permitting.

These vast distances were covered at speed for the purposes of colonisation or commercial gain but many riders find an inner spiritual wealth to be gained from equestrian travel where speed and distance are not the priority. Long Riding today is not concerned with vested interests and setting records but with a life that embraces new horizons and is lived at a different pace and with different values.

'Nature did not design horses to live like rats in small stalls. They, like humans, are born travellers. Thus this unique combination of equine muscle and human brain created a centaur capable of exploring the planet together.' These are the words of CuChullaine O'Reilly who founded The Long Riders' Guild in 1994 to collect and make available the international expertise and experience of equestrian travel.

Membership of the Guild is by invitation and qualification to become a Long Rider requires a documented, continuous ride of a thousand miles or more, during which the welfare of the horse has been paramount. There is no competitive element whatsoever in terms of speed or distance and there are no prizes. There are no restrictions on the type of horse or the rider's age, gender or nationality but he or she must be competent to ride and to look after every aspect of the horse's physical and psychological well-being.

The most famous 20th century Long Rider of all is Aimé Tschiffely (1895-1954) of Swiss nationality. In 1925, finding his teaching job in Buenos Aires rather dull, he decided to ride from there to New York. He chose to take two horses, aged fifteen and sixteen and only roughly broken in. Both were of the native Argentinian breed, the Criollo, known for its toughness. His solo journey of over 10,000 miles took two and a half years through deserts, coastal jungle and the Andean mountain ranges.

After extraordinary adventures, Tschiffely and the horses (Gato and Mancha) arrived in New York and were given a hero's welcome with a ticker-tape parade. He met President Coolidge and was invited to give talks about his journey to various societies while the horses were exhibited at Madison Square Garden.

Tschiffely planned to return the horses to Argentina where they could enjoy a well-deserved retirement on their native pampas so he booked a sea passage for the three of them to Buenos Aires on the Vestris. However, the National Geographic Society in Washington persuaded him to give one last lecture and he postponed their departure. This invitation probably saved the lives of all three of them; the Vestris was shipwrecked and over a hundred passengers drowned. Both horses returned later with him to their native pampas and lived to the great ages of thirty-two (Gato) and thirty-seven (Mancha).

Tschiffely's enormous influence continues to inspire many Long Riders and countless others today. He is remembered in a law passed by the Argentinian Congress to celebrate September 20th, the day of his arrival in New York in 1928, as their annual National Day of the Horse. He died aged fifty-nine from complications following a minor operation. His ashes were scattered on the pampas near a stone memorial to Mancha and Gato which was erected later on the El Cardal ranch in Argentina.

Tschiffely's achievement was probably largely due to his meticulous attention to detail in the planning of his journey, his patient and determined mindset and the priority he gave to his horses' well-being.

Preparation is vital. Long rides have an irresistibly romantic aura but riding into the sunset far away from the cares of the world without a great deal of preparation is to dangerously disregard the practicalities. A modern Long Rider, William Reddaway, travelled over 2600 miles visiting thirty cathedrals in England on his horse, Strider, to raise

Gato being led over a Peruvian chasm on a swinging cable bridge.

money for charity. He found that it took over six hundred hours to arrange accommodation and provisions for them both and three hundred more hours to work out the best route (using 125 different maps!). That was apart from the time taken to get themselves fit and for him to buy the right kit.

Riding and wild-camping (as described by Richard Barnes in *Eye on the Hill*) is the alternative to organised accommodation but its unpredictability means that extra supplies for rider and horse must be carried.

Time for the training of a packhorse is also important as not all horses turn out to be temperamentally suitable. The ideal horse would be a stoical gelding which can keep up with the pace, follow reliably and stand quietly while being loaded. It must learn to accommodate the dimensions of the pack, particularly alongside its road horse companion and near walls or trees. Inevitably there will be road work so all horses should be reliable with traffic.

Aimé Tschiffely's Mancha and Gato resting in a dry river bed in southern Bolivia.

Most of us ride on the same horse for only a few hours a day. A long ride, where the horse may be ridden for six to eight hours a day for weeks, months or even years, and is rested for two days in every seven, is quite a different matter.

Equestrian travellers don't have tack rooms, not even the portable caravan of the circus or the international competitors' horsebox. Theirs is a minimalist, movable, highly selected collection of tack, perhaps the essence of a tack room taken beyond walls. Every strap, brush, and spare horseshoe nail has to earn its keep to qualify for the journey. Improvisation is helpful too: dental floss is strong enough for temporary tack repairs as well as good for sewing up rips in a tent or clothing, quite apart from keeping the dentist at bay.

The most important piece of kit for a Long Rider is the saddle. Constant wear on the delicate skin and anatomy of a horse's back can quickly cause damage and saddle sores which may take weeks to heal. Pressure points caused by a poorly fitting saddle can result in bruising or vasoconstriction in the muscles and a painful back can lead to lameness. A pack saddle is a dead weight which does not shift so is harder for the horse to carry and must also be adjusted with the greatest care.

Many long-distance riders favour an Australian stock saddle. It is relatively light (about 12lb), has long panels like a cavalry saddle to distribute the weight, knee rolls to stabilise the rider and plenty of attachment points for equipment. A modern saddle designed in France, the 'Randonnée', is also a popular design. It has a generous arch in the front allowing free movement of the horse's shoulders and wide side panels. The leather underside is padded inside with thick felt and a high pommel and cantle create a secure seat for the rider. Len Brown developed the Ortho-Flex saddle from his experience of riding three thousand miles in the US and his saddles were used successfully for a 19,000 mile ride from Patagonia to Alaska. Since then he has developed felt saddle pads, adjustable by adding shims (extra pads): the Protector and Corrector designs, which can be used under any saddle.

A Long Rider may also walk many miles leading his horse so will be wearing good walking boots. For this reason the stirrups must have plenty of width to accommodate a wide foot.

Treeless saddles, while comfortable for the rider, do not protect the horse's back as well as a properly fitting one with a tree. This was the conclusion reached in 2008 by the Society of Master Saddlers who carried out the comparison. Newer designs of treeless saddles have improved the protection of the spine and the flow of air over it with firmer padding but they are definitely not recommended for the rigours of long-distance rides.

The saddle pad is crucially important to the success of the journey. Anything that touches the horse's skin has the potential to cause friction resulting in a serious saddle sore. Long Rider Sabine Keller has ridden across Europe twice and, instead of a packhorse, she uses Ortlieb saddlebags attached to her riding saddle. These are soft, waterproof bags which have been used successfully by several Long Riders over a combined distance of at least 20,000 miles. Underneath the saddle is a thick longwool sheepskin pad (wool side to the horse) which is large enough to prevent any equipment actually touching the horse's back. On top of that and under the saddle is a second pad of felt.

This combination was comfortable for horse and rider so long as the sheepskin was dried and brushed out regularly and the felt, which becomes hard when wet, did not contact the horse. As a rough guide to weights, Sabine's bags weighed about fifty pounds, the saddle was twenty five pounds and with her own weight the total was about two hundred pounds.

An interesting saddle blanket made of woven horsehair, thought to have ancient Mongol and Siberian origins, was described by Tim Cope during his ride across the Asian steppes. The advantage of horsehair is that it dries very quickly after becoming wet from rain or fording rivers and is much less likely to rub the horse's back.

For pack saddles, Long Riders recommend the Canadian model made by Custom Pack Rigging. This is an adjustable framework of steel or aluminium and plastic which can fit any size of mule, horse or even camel. Lockable, waterproof plastic boxes, which are concave to fit the ribcage of the animal, can be

William Reddaway on Strider completing his ride round thirty English cathedrals and over 2600 miles from 10th May to 8th December, 2013. His saddlebags are much lighter than when he set off having reduced his equipment to the bare essentials.

attached in various configurations and used as stools or tables in camp.

The decision to take a packhorse or not depends on many factors, such as the availability of food and feed and the amount of road work and traffic. Having two horses obviously doubles the rider's responsibility for their well-being but allows a more flexible schedule – supplies for all, including camping gear, can be carried for a few days without having to find daily provisions.

If you choose to take a packhorse, it is helpful if the horses are similarly paced and also like each other. On a long journey with two of us riding borrowed horses, it had not occurred to me that they could be arch enemies; an oversight which nearly ended in disaster. Both had come from the same trekking centre and while on the road appeared to tolerate each other. At the end of the first day they were both turned out into a small paddock with plenty of grass. Immediately, one attacked the other with teeth and heels flying and would have inflicted serious injury if we had not managed to separate them. The owner of the only cottage in sight was surprised but agreeable to our urgent request to allow one horse to spend the night in her front garden. After the horses had been individually settled, she then settled us too by inviting us in and pouring us each a double gin and tonic. How remarkable is the kindness of strangers!

It is not the kilometres as much as the kilograms which are a major factor affecting the horse's performance. So many wise sayings from different sources concur on the importance of the lightest weight possible to be carried by either a road horse or a packhorse.

The rider's own weight and pack are the main concern but the horse's shoes should also be considered. Using heavy-duty shoes for a lot of road work may be a false economy. Medieval Islamic texts cautioned that 1lb on the foot was equivalent to 8lb carried on the back.

An average European horseshoe typically weighs up to 2lb but a 'town shoe' may be much heavier. This can be significant. We bought a sturdy grey pony in the early 1960s at auction for £25 without prior examination. We took the pony home and after it had rested I rode it to see what it could do. It willingly approached the 1ft high pole I asked it to jump but knocked it down each time. It had been shod with the heaviest town shoes – for pulling a cart on the streets of Liverpool, as we later discovered – and when these were removed the pony was transformed into a careful and competent jumper.

The problem of weight and wear can be helped by using tungsten headed nails in regular shoes or by welding on patches of borium (tungsten carbide crystals) to steel shoes during their manufacture. The use of Easy Boots (supplied to Long Riders Louis Bruhnke and Vladimir Fissenko for their nineteen thousand mile ride from Patagonia to Alaska) is an alternative or just shoeing the front feet which carry most of the weight.

A long-distance rider, having overcome the inevitable prophesies of doom from colleagues, may then go through many different states of mind. On the first day he may be thinking of the home he has left and about how his loved ones will manage in his absence. There will be worries about what may lie ahead and whether his preparation has been sufficient. Being cut loose from the umbilical cord of the internet, familiar routines and company, requires adjustment.

The next stage might be accommodation to the pace; experienced riders only walk their horses so until the cadence becomes habitual and enjoyable, progress may seem slow. Road horses may very occasionally trot but packhorses should never do so. Taking the time to savour the texture and history of the land, be out in the weather and to be at ease with

Solas the Shire and Skye the Clydesdale power up to the top of the Cumbrian mountains.

Cumbrian Heavy Horses 2017. Pure bred Ardennes, Clydesdale, Shire and Brabant enjoying a deep wade off the Cumbrian coast.

solitude are just some of the joys of long riders. On other days if cold, hungry and tired, self-reliance and resolve will be tested. But in the judgement of the true devotee, resilience will always be rewarded and who could ask for a better long-distance companion than your chosen horse?

While progress may only be about three or four miles an hour it is possible to achieve fifteen miles a day more or less, depending on the terrain. A horse walking at this pace does not seem to frighten wild life unduly and it is a special pleasure to see birds and mammals at close quarters. Daily mileage on such expeditions used to be higher when the population was more used to seeing horses. A Long Rider these days creates such a public interest that a significant amount of time is often spent in explaining about the journey to the small crowds and local press who gather, especially on urban routes.

But what of the horse's state of mind? We can only imagine his response to being taken out of his familiar environment and made to spend each night in a different, potentially hostile place. Evidence suggests that the horse will become dependent on his rider and a bond of trust will develop. The rider is the one constant in his life, his guardian and provider.

It has been my experience, as well as that of others, that once the horse has recovered from being parted from home, enthusiasm for going forward into new territory with pricked ears and a spring in the step takes over. It feels as though a fragment of the equine DNA that urged the primitive horse on the steppe to keep roaming in search of water and fresh grazing, has been awakened.

Experienced riders claim to discover an elevated mental plane in which they do not dwell either on what lies behind them or on their destination ahead. Their geographical progress is a physical journey but for many it is no more important than their cerebral journey. For them, the saddle is now home wherever they are. They have achieved a quiet ecstasy, a fulfilment, an uncluttered and creative mind and new levels of awareness. They relish the shedding of the many small millstones which grind through normal humdrum life, content in the freedom of their life on the road and whatever it may bring.

This precious state of freedom and excitement has been eloquently expressed by riders of many faiths and languages through the centuries; a Taoist saying 'The journey is the reward' was endorsed by

Linda Klasing, a German student on a summer placement with Annie Rose at Cumbria Heavy Horses. Most of the heavy horses here fit a regular width saddle used with a saddle pad and neoprene girths which are soft, stretchy and can easily be sponged clean. Breast plates prevent the saddle slipping on the steep gradients and knee rolls steady the rider.

Robert Louis Stevenson's quote 'It is better to travel hopefully than to arrive' after his journey through the Cevennes with a donkey named Modestine. Perhaps the best of all is an old Arab description, 'The air of heaven is that which blows between a horse's ears'.

Long Rides

Most of us do not have the opportunity to make journeys on the grand scale described by the Long Riders, but can enjoy the experience of rides for a day or week at a time.

The Cumbrian Heavy Horses company specialises in bespoke rides for individuals or groups in the beautiful countryside of the English Lake District. This is different from trekking where generally customers have to fit in with the routes and horses available. Here, riders are carefully matched to the temperament of the horse and the type and length of routes are mutually agreed. They are accompanied by staff trained at the Mountain Rescue Centre to deal with every situation.

I visited the traditional stone farm on a glorious autumn day and was greeted by about a dozen heavy horses; The Centre has several different breeds of purebred heavy horses from Shire and Clydesdale to less common Suffolk Punch, Ardennes and Brabants, all relaxed and snoozing together in the open-sided barn. The tack room was being aired and tidied in the sunshine by students, watched by a collie and three cats.

Cumbrian Heavy Horses is unique. Annie

Rose, the director, has been largely responsible for promoting the heavy breeds as excellent riding horses and rescuing them from obscurity. Her horses prove all the sceptics wrong; Clydesdales – in fact all of these big horses – are not wide but relatively narrow and use regular sized saddles. They are not boring and slow to ride but forward-going and lively and love a gallop along the beach. Their height (17-18hh) gives the rider a commanding view and their natural power provides a Rolls-Royce ride. Riding a heavy horse is a very special experience and I thoroughly recommend it.

No detail escapes Annie's attention and her reputation for horsemanship and management is such that foreign and British students come here to learn during the summer months. When she isn't riding or supervising the yard she is writing articles on heavy horse matters, arranging events, developing the stock programme and running the small campsite on the fell farm where the Centre is established.

The tack room is in a 150-year-old stone barn and runs according to strict rules. Saddles on racks, each labelled with a horse's name, have a code; if they need cleaning they are placed with their cantles against the wall. Cleaned ones are placed the opposite way round with the cantle outwards. The saddles are 'Free'n Easy' made by Les Sparks at Barnard Castle. They have long panels to distribute the rider's weight and knee rolls to keep him steady on the steep hills. Bridles that need cleaning are hung on a low hook next to the saddle and moved to an upper hook when clean – a simple and efficient system which everyone follows.

In a barn next door there are a number of rugs draped over poles to dry. Each pole has the horse's name painted on the end so nothing gets muddled up. In the feed room everything is scrupulous with the aroma of flaked maize and barley, rolled oats and sugar beet. Limestone flour can be added to the feed of young stock to provide calcium and for any older horses there is cider vinegar to ease their joints.

Originally based on Skye, Annie and friends rode all her horses from there to their present base in the Whicham Valley, Cumbria in 2006. Here there is better grazing, a longer season, easier accessibility and wonderful beach and hill riding. About twenty lucky heavy horses live and work here. I have never seen horses so nearly smiling!

The popularity of the heavy horse is increasing in many directions, as shown by the variety of carriage and ridden classes at the annual National Shire Horse Show held in Stafford. These horses are fun because they are so adaptable: the Household Cavalry's drum horses carry the solid silver kettle drums on ceremonial occasions and in Scotland, Clydesdale racing has been held at Irvines Marymass Parade for decades to test their fitness.

In the woods and fields heavy horses are extracting timber and ploughing and in Cumbria they are providing superb off-road riding. In the show ring, one of the most enduring images I have is of a champion Shire horse giving a gloriously exuberant dressage demonstration – he floated across the arena in a spectacular extended trot with all feathers flying bringing the spectators to their feet in a standing ovation.

Competitive off-road riding organised by Trec GB may appeal to some riders. Trec is originally a French driving discipline which was adapted as an orienteering competition for Britain in 1998. It is designed to test skill in planning and executing a long ride in unfamiliar country. Trec events usually have a presentation check followed by cross-country orienteering, a 'control of paces' phase and an arena obstacle course. It is ideal for novices as well as experienced ponies, horses and riders of all ages.

There is a good choice of bridleways in most areas of Britain for independent long-distance rides. Access in England is not as easy as in Scotland but is improving all the time. Riders can speak to landowners in advance and many are sympathetic. Old railway lines and canal towpaths are worth reconnoitring and Roman roads, historic droving and packhorse routes or parts of General Wade's Scottish military roads built in the Jacobean period, give wonderful riding experiences.

The Buccleuch Country Ride in the Scottish Borders is a 59-mile, waymarked ride, off-road or on quiet country lanes. In Wales, the Great Red Dragon Ride is a combination of five different British Horse Society-approved rides totalling nearly three hundred miles from Prestatyn in the north to Port

Left to right: Miracle the Shire, Ben the Clydesdale, Eddie the Shire and Skye the Clydesdale in 2017. Saddle bags made of tough canvas with reflective strips can be attached to the saddle for picnics during longer rides.

Talbot in the south.

There are plans to extend the Pennine Bridleway National Trail, a route already over two hundred miles long, linking Yorkshire, Cumbria and Derbyshire. The Sabrina Way, opened by HRH The Princess Royal on horseback in 2002, links the southern end of the Pennine Bridleway to the Cotswold Way and many other routes.

Long Rider Vyv Wood-Gee recently designed and completed her own 2400 kilometre ride in Britain to raise money for cancer charities. Her route connected eight white horses carved into the country's chalky hillsides and a landscape sculpture of Sultan the pit pony (*see* page 163) created on a coal tip in South Wales. She began her three and a half month journey at the Mormond white horse near Fraserburgh, Scotland and finished at the white horse in Uffington, Wiltshire.

These rides offer a non-competitive adventure in some of the most beautiful scenery in the British Isles and information about routes and nearby accommodation for the horse and rider can be obtained from the British Horse Society.

CHAPTER TWENTY-THREE

Mounted Police

Before the introduction of the motor car, horse power was essential for public services: prison vans, ambulances and fire-engines were all horse-drawn. The first provincial mounted section was formed in Liverpool in 1886 and was then chiefly concerned with transport. It is said that the fire service horses were so well trained that when the fire bell sounded and the horses were released from their stalls they assembled themselves in the correct order in front of the fire engine. The harness, which was suspended from the roof above, was automatically lowered into place, buckled up and the engine was ready to move out within minutes.

This was only marginally slower than the modern fire engine today which has the advantage of an ignition key.

Full collars were used by the two or three horses needed to pull the considerable weight of the fire engine and its tanks of water, but no breeching was included in the harness as the engines were fitted with hydraulic brakes. The collar of a fire-engine horse is now a rare item but is instantly recognisable by its latch fastening under the neck which took only moments to secure.

In 1938 the Section moved from Liverpool city centre to purpose-built premises in the suburb of Allerton and today some twenty-four horses are stabled there and still train in the original indoor school. During the war, the horses were moved out to Knowsley Hall by Lord Derby, away from the bombing raids on Merseyside.

The tack room is panelled and hung with gleaming saddles and bridles of English leather.

Everything is neat and tidy, ready for routine training and patrols or to spring into action for an emergency if necessary.

The original mobile cast-iron saddle horses are still in use and a high shelf running round the room holds at least a dozen silver cups won at police horse shows. Every year a public event, the Bluelight Horse of the Year Show, is held in which horses from the police and fire services all over the country compete in dressage, show jumping and obstacle courses.

Bridles are fitted with Cavalry Pelhams which give extra control, as do the spurs worn by the riders, when needed. Leather saddle bags can be attached to the saddle containing equipment needed for patrol duty.

The most vulnerable parts of the horse are the bony areas of the head and legs. For riot control, the face and eyes are protected with a clear perspex eyeshield and a leather guard runs from brow to muzzle while padded shin and kneepads are worn on the legs. This gives the protection and flexibility needed for the crowd control strategy: to move in fast to break up the crowd and create a space and then to retreat as soon as police back-up arrives.

High visibility is an important and reassuring part of the mounted police presence in a large crowd. At the same time police riders have a view over the heads of the crowd and can relay information quickly. Cameras in the riders' helmets and radio links allow the control centre to keep abreast of rapidly changing situations.

Liverpool City mounted police on routine patrol. The horses are wearing protective Perspex face shields and the police riders' helmets are equipped with cameras.

Horses come from a variety of backgrounds. Murphy's Law, for example, is an imposing black horse who had already had a chequered career. He arrived as a youngster having fallen into a septic tank and been rescued by the fire brigade. After that he was sold at Beeston Market and then resold to the Mounted Police. The ideal mare or gelding is typically Irish Draught, standing 17hh and with a sound constitution for a service life of about fifteen years.

Training is tailored to each horse and may take from twelve to eighteen months depending on its character. The horses always work in pairs as they are herd animals and a youngster is paired with an older more experienced horse.

The 'Street Nuisance' training is designed to prepare the horse for any eventuality it may meet while on duty. An obstacle course is set up in the indoor school and in this controlled environment, horses learn to cope with loud noises, flags and missiles at close quarters and a variety of surfaces underfoot.

Watching a novice horse tackle such hazards one can see the conflict between obeying the rider and the instinctive impulse to flee, but eventually trust and experience triumph. Schooling includes training the horse to move sideways and backwards at speed and to withstand the press of a crowd.

The horses have varied duties apart from patrolling the Liverpool city streets and the sand hills and pine woods of Southport and Formby. They are particularly good for public relations with the community and carry out special duties at football matches and Aintree Racecourse. Mounted police are routinely on duty in Liverpool city centre throughout the night every Friday and Saturday.

Police riders are matched to horses depending on their ability after a sixteen week intensive riding course. They remain with their horses for twelve to eighteen months to develop the trust and respect needed for effective policing, carrying out all the grooming to help bond with the horse.

One experienced rider and a well-trained horse is reckoned to be equivalent to ten policeman on the ground so they are highly cost-effective. They are superior to the police car for public relations, in overall cost and maintenance and very often in crowd control too.

When Liverpool won the European Cup in 2005 the crowd thronging the streets to welcome back the team was at least 10,000 strong. At one point near St George's Hall, the team bus, escorted by police horses, was blocked by the crowd and forced to stop. Sixteen police horses formed an arrowhead formation in front of the bus, parting the crowd like the Red Sea so the bus was able to proceed. The horses were quite oblivious to the confetti cannons going off all round them, the roar of the crowd and the shattered champagne bottles and beer cans beneath their feet.

It is a tribute to the benign willingness of the horse to cooperate with man, as well as to the training, that instinctive flight in the face of aggression can be overcome. The flexibility and latent power of a police horse ensure that it remains one of the most effective forms of crowd control in the world, despite the array of modern technologies.

Each horse's equipment is arranged above the riot gear pack which is neatly wrapped in high-visibility horse sheets on the shelves below.

CHAPTER TWENTY-FOUR

Pit Ponies

Few of us know the absolute darkness of a mine. It can be experienced on a tour of Big Pit, the deep mine at Blaenafon in South Wales, when the guide turns off his head torch. For a long two minutes, enveloped in blackness, hundreds of metres below ground with the occasional dripping sound of water and the smell of coal, it is possible to guess at the working conditions for the miners and seventy resident horses who worked here. Then there would have been more noise: men shouting, the bang of ventilation doors opening and shutting to let the horses through with their loads, the grinding of metal wheels on rails and the faint glimmer of a miner's lamp. A more hostile environment is hard to imagine; the miners had little choice of work and the horses had none.

Most horses which worked underground were employed in coal mines but some were also used in the Cheshire salt mines, the Yorkshire Dales lead mines and the Welsh slate mines. These last three could be classified as 'drift' mines where the access is an inclined passage down to the face and the horses were stabled above ground overnight. The deep mines could only be accessed by vertical shafts with a cage lift where the horses were stabled in stalls underground, often for years.

Carl was a 12.2hh Welsh Mountain pony who was one of the last working ponies in the deep mines in England. He retired in 1994 from the Ellington Colliery, Northumberland after twelve years underground, aged seventeen. During the day he worked six miles out under the North Sea and at the end of his shift returned to his underground stall. Carl represented pit ponies at the Queen's Golden Jubilee in 2002, appearing in 'All the Queen's Horses' and spent his retirement at the National Coal Mining Museum in Yorkshire taking part in events and local shows.

At the height of production, eighty ponies were employed at Ellington and accommodated in stalls built four miles into the mine from the access shaft. The ponies had a fortnight's holiday above ground every year but this was a controversial issue. During the fortnight, the ponies often damaged each other while establishing a natural pecking order and the change in diet and exposure to the weather after the constant temperature in the mine could make them ill. Having tasted liberty, they were reluctant to return to life underground.

The geology of the coal seams determined whether horses or ponies were used. In South Wales the coal could be hewn out in large blocks and loaded into trams (drams or tubs) which weighed two tons each and needed a strong Welsh cob of about 15hh to move them.

In the Durham coalfield, seams were often so thin that the working height at the coalface could be as little as 2' 6" and the miner would be either on hands and knees or lying sideways while wielding his pick. Shetlands of 10hh were widely used as well as the bigger Galloway ponies, bred for the purpose in Scotland. These 'Gallowas' as they were known in pitmatic, the patois of the north-east mining communities, were the most popular being very sturdy with calm temperaments. Large draught horses were also used at the surface either pulling a series of loaded tubs or walking in a circle harnessed

Little Tick, a favourite Shetland pony working in the Durham coal mines in 1913.

to a gin which operated machinery below ground.

There are many stories of ponies and their drivers developing a strong bond. For a driver attuned to his horse, paying attention to any unusual behaviour often saved both their lives. A number of reports describe the sudden refusal of a willing horse to go ahead having detected movement in the strata above and a deadly roof fall happening shortly after.

In the first half of the 20th century demand for coal was so high that extra horses were imported from America, Canada and from as far away as Russia to maintain the horsepower in the British mines. Ceri Thompson, an authority on Welsh pit ponies, told me that even a Przewalski's horse had somehow finished up working in Big Pit, a bizarre end for that most feral of horses.

The mines throughout the country varied greatly in the standards of care for their human and equine employees and until the late 19th century there was little welfare protection for either, though women and children under ten were banned from underground work in 1840.

In the 1790s, cast-iron rails were introduced and the number of pit ponies escalated rapidly. The first legislation to protect them was not until nearly a hundred years later in 1887 when inspectors could make limited investigations into how the ponies were treated. The treatment of all animals in Britain was regulated by the Protection of Animals Act in 1911. This was modified in 1949 and 1956 to become 'The Pit Ponies Charter' which significantly improved the living and working conditions of ponies underground. Nationalisation of the pits into the National Coal Board in 1947 had also raised standards; shoeing had to be done by qualified farriers and the first horse hospital was built at Tondu, Glamorganshire, for the four hundred local working pit ponies. Their injuries were almost entirely physical and the coal dust pulmonary diseases, such as pneumoconiosis which affected the miners, do not seem to have been a notable problem in the ponies. It is a myth that working in near darkness caused blindness although their eyes took time to accommodate to daylight above ground. Ponies blinded through injury were a danger to others and were immediately dispatched.

The Charter decreed that ponies could not start work in the mines until they were four years old, daily records of their shift work had to be kept and protective headgear and eye guards were compulsory. Each pony was in the care of two people: a horsekeeper who looked after the feed, water and sawdust bedding for every fifteen ponies and a driver who was responsible for his pony while they worked.

When the pony was returned to the stall, the horsekeeper would check the pony for injuries, its shoes and harness. If there were any scratches on its leather back protector, indicating that a low roof had caused the abrasion, the pony was not allowed to continue working until the passage floor had been dug out.

The number of pit ponies declined as they were made redundant by machinery or pit closures. In 1878 it was estimated by the RSPCA that 200,000 ponies worked in British mines declining to about 70,000 in 1913, 2,000 in 1969 and to 55 in the 1980s.

There are almost no memorials to the pit ponies, apart from the magnificent earth sculpture by Mike Petts. It depicts 'Sultan' a pit pony known in the local folklore for his character. He would steal the miners' sandwiches from their jacket pockets and grab any tea canisters he could find, remove the stopper with his teeth and upend the contents into his mouth while evading capture. During the bureaucratic reorganisation of the National Coal

Board to the British Coal Corporation in 1987, many of the mine records in Cardiff were destroyed so Sultan's dates are unknown but he probably lived around 1900.

It was coal that fired the Industrial Revolution which propelled Britain to becoming a world power. It was the hard, top quality coal from the deep Rhondda mines which produced temperatures high enough for the Royal Navy engines to run optimally during World War II. Once called black diamond because of the personal fortunes created in the industry, fossil fuel is now ecologically shunned.

The coal era, the millions of miners and the pit ponies who provided heat, light and power have become part of our history.

Underground stabling in Lambton D pit, Durham, in 1920. The stalls were lime washed regularly to promote hygiene. Eight national inspectors were employed to examine about 68,000 horses at this time.

Ton Pentre, S. Wales. The harness is typical: heavy-duty collar, pad and breeching connected to the front end of the 'limber', a steel rail running round the pony's hindquarters to which the tram was attached at the rear. Note the leather plate protecting the pony's hindquarters from striking the roof. Date unknown but before 1949 as there is no eye protection on the bridle.
There are no bits, just a single rein for the driver to guide the pony from his perch on the front of the tram.

Steel in retirement with Roy Peckham who founded the Pit Pony Sanctuary near Pontypridd. Steel took part in the Horse of the Year Show parade of 'Horses through the Ages' where his astonished neighbours, the Household Cavalry, learnt that he could pull ten tons of coal.

Steel, one of the last ponies to work in a Welsh drift mine in 1999. Each of the loaded trams weighed two tons. Steel refused to return the empty trams to the pit until he had had a polo mint and consumed a packet a day.

A 200-metre sculpture made of coal shale at Penallta, Caerphilly. This millennium project is a memorial to the pit ponies of South Wales and shows Sultan galloping free in the fresh air and sunlight.

CHAPTER TWENTY-FIVE

Polo and Polocrosse

The sound of a polo mallet hitting a ball is nearly as English as village cricket. Cricket, though, started in the 16th century in south-east England while polo worked its way westward becoming a British Army game only in the 19th century. Its popularity has spread quickly and worldwide. It is now played at all levels by a wide range of riders from top professionals to Pony Club members and in Argentina at the Potrillos Cup tournament for the under-14s, there are teams made up of those aged under eight years old.

Polo has a long history which is thought to stretch back over two thousand years to its origins in Asia. Known as the 'Game of Kings' because of its close association with royalty and cavalry training from ancient times, polo is depicted in paintings of the Chinese Tang dynasty (618–907AD) and in the 14th century Persian Shahnameh ballads.

British polo has its roots in the aromatic earth of Northern India and Pakistan where Gilgit and Manipur have the oldest modern polo grounds dating from the 1860s. Army officers stationed in India observed the game and eventually a form of it (first played with walking sticks and golf clubs and a cricket or billiard ball) arrived in England. Polo became established at the Hurlingham Club, a fine Georgian house on the banks of the Thames, where it out-competed the Pigeon Shooters' Society (the spectators far preferred polo to shooting), though to this day the pigeon remains as the Club's emblem. The Hurlingham Polo Association published the rules of modern polo (still followed worldwide) in 1873 and the first game between the 1st Life guards and the Royal Horse Guards and the Lancers was held in 1874 with royalty among the spectators. Polo became an Olympic sport in 1900 until 1936. During the Second World War the Hurlingham polo grounds were requisitioned; the RAF placed a balloon barrage on the No. 1 field and the No. 2 field was used for an anti-aircraft battery.

Polo is no longer played at Hurlingham though its origins are cherished and many fine paintings (particularly by Gilbert Holiday, 1879–1937), trophies and bronze sculptures are on display. The most important bronze is 'The Big Four' by Hazeltine which was presented by Harry Payne Whitney's American team following victory in 1909 for the Westchester Cup, a contest which continues to the present day. The leading British polo clubs are now the Guards' Polo Club in Windsor Great Park, Cowdray Park and Cirencester Park.

Argentina first became involved with British polo apparently through the Correspondence page in *The Field* magazine in 1870. In reply to an enquiry from a British settler in Argentina as to whether it would be a suitable game for that country, another correspondent wrote in recommending it and advised that plenty of brandy and soda should be provided for the players. With its favourable climate and terrain, ranching culture and the native Criollo breed, Argentina took to polo like a horse to its oats. Today more than half the high-goal players in the world are Argentinians. The highest rated individual is Adolfo Cambiaso. He is one of a handful of players with the maximum 10-goal handicap, which he acquired at the age of nineteen and he has broken more records and won more trophies than anyone else.

A Persian miniature illustration from the Epic of Shahnameh by the poet Ferdowsi, circa 1340.

From being exclusively a game played by army officers to hone their cavalry skills and prevent boredom during long tours of duty abroad, it became a favourite of the aristocracy and wealthy amateurs. The Duke of Edinburgh, Prince Charles, Sir Winston Churchill and the 3rd Viscount Cowdray all played high-goal (highest standard) polo. Princes William and Harry were taught by the legendary Carlos Gracida, a 10-goal player from Mexico.

Professional and highly glamorous polo has now evolved in which patrons finance hand-picked teams which travel round the world competing against each other in high-goal tournaments. This is exhilarating polo – fast, highly skilled and not without risk.

It is a game of maximum adrenalin, a high speed, contact sport, gladiatorial yet graceful and balletic at the same time. With their speed and agility these ponies could also excel at flat racing or dressage with pirouettes on the haunches, flying changes and half halts from a gallop. Their turning, braking and acceleration are spectacular. A top polo pony can be worth £100,000 or more and each player would expect to take five or more ponies to a high-goal tournament, one or even two for each chukka (period of the game).

A large number of top polo ponies are mares who appear to relish the body contact required for 'riding off' their opponent. There are good geldings too and occasionally stallions but this bias towards mares may be because in a wild herd it is the mares who are the disciplinarians. They drive out badly-behaving colts from the group with their teeth and heels, isolating them for a few days before allowing them back into the herd. The most important attributes for a polo pony are a good mouth, a quiet temperament, good balance and plenty of speed and courage.

As polo has become ever more competitive, thoroughbred ponies are preferred for their speed. Jack Richardson owns about thirty ex-racehorses which he is retraining for polo, twelve of whom currently play in his main string. One of them, Harmony World who ran on the flat six times,

The Snow Polo Cartier World Cup in St Moritz, 2017 on the frozen lake. Perrier-Jouet v Maserati. The old height rule of 14.2hh for polo ponies was abolished in 1919 and ponies are usually 15–16hh today.

won the 2017 Retraining of Racehorses Elite Polo Champion Award.

Many patrons and players have their own breeding operations and a controversial development in the production of polo ponies is well advanced in Argentina. Adolfo Cambiaso has pioneered cloning techniques and regularly plays on cloned ponies in high-goal tournaments. In 2016, he played a different clone of Cuartetera (a favourite mare) in each of six chukkas at the Tortugas open final and now his entire team of ponies is often cloned.

In 2006 his world famous stallion, Aiken Cura, was seriously injured during a tournament and had to be euthanized in 2007. A small biopsy, taken from the neck and containing Aiken Cura's cells, was frozen and stored. From this a number of cloned ponies were developed by Crestview Genetics Argentina SRL, founded in 2009 by Adolfo Cambiaso, Ernesto Gutierrez and Alan Meeker. These cloned ponies are apparently normal animals, 85% of which perform as well as, or better than expected, on the polo field. As a rough estimation, genes may account for 30% of the animal's likeness to its 'parent' while 70% is accounted for by diet, training and its environment. A three month old clone of Cambiaso's horse, Cuartetera, was bought for $800,000 at auction, a record price for a polo pony sale.

Snow polo is a spectacular variety of polo and its success has spread round the world. It is now played in Moscow, Beijing and Aspen, Colorado. St Moritz was the first venue to introduce snow polo in 1985, though grass polo had been practised there by the British Cavalry as long ago as 1899. The first horse race on the frozen lake was in 1907.

Rudyard Kipling, a keen polo player, spent the winter in St Moritz from 1909 to 1913. In his jewel of a short story *The Maltese Cat* he analyses a thrilling game of polo in India, narrated from the ponies' points of view. Kipling makes the interesting claim that a good polo pony understands the strategy of the game so completely that it is not necessary for it to wear a bridle during play. It is known that the best ponies learn to chase the ball but few players would dare to test his claim!

I was a spectator in January 2017 at the Snow Polo World Cup in St Moritz where polo is played on the frozen lake. The ice was half a metre thick, strong enough to support the Polo Village, ponies, cars and the landing of light aircraft. It was a magical scene: the Engadin Valley stretching away between high peaks, sun on snow and the vibrant colours of the polo festival on the ice. Live music, champagne and plenty of fur-clad spectators all added to the excitement.

Four international teams play each other in a number of games over three days and the prestigious Cartier trophy is awarded annually to the highest scoring team. The Cartier v Maserati game was particularly exhilarating. As I stood watching, several riders at full gallop came straight towards me and I felt the impact of a wave of high energy. The power and awe of this cavalry charge was electrifying. The ball banged hard into the barrier immediately in front of me and an instant later it was hidden in a whirling skirmish of riders and ponies, legs and sticks amid sprays of glittering snow. Then with a shout the orange ball was sent flying towards the goal, the teams wheeled like quicksilver and raced away.

For the world polo fan, the glühwein and white turf in Switzerland can be exchanged for Pol Roger at the beach polo at Glenelg, Australia, a Pimms or British Polo Gin on the green turf of England, cocktails at the palm-fringed polo fields in Florida or an iced Fernet at the iconic Palermo Field 1 in Argentina – on to India, Thailand and many more venues.

Horseshoes worn for snow polo have a stud in each heel for grip and an inner filet of rubber tubing which prevents the snow balling in the hoof. There is a covering of a few inches of snow on top of the ice which gives the ponies a surprisingly good grip.

The Black Bears Polo Team

The Black Bears is a high-goal team, founded by Urs Schwarzenbach, based in England and Australia. His son, Guy, has taken the reins and is now the Patron and an active player. Polo is a passion, often shared within families, such as the three Argentinian Pieres brothers and their cousin, which can give teamwork an advantage.

Behind the scenes of this international high-life is a superbly organised operation: breeding, schooling, practice, more practice and a good deal of success in all the major polo tournaments.

The Black Bears' base in Oxfordshire is a fine estate sloping down to the wooded banks of the Thames with five perfect grass and all-weather polo grounds. I am shown round by Urs' polo manager, ex-Household Cavalry riding master, Douglas McGregor. We enter the first stable yard through an ancient listed timber-framed barn built from ships' timbers at least three hundred years old, a complete contrast to the modern stable yards beyond. About a hundred looseboxes surround the three yards, attractively built in brick and local flint with red tiled roofs. All the walkways are rubber-paved, quiet and non-slip.

Heavyweight leather for all polo tack: breast strap to secure the saddle, draw reins for extra stopping power, a double girth and a standing martingale to prevent the rider's face colliding with the back of the horse's head when leaning forward to hit the ball.

Urs Schwarzenbach (left) patron of the Black Bears high-goal team, discussing sticks. The shaft is made from manau cane which is not hollow like bamboo but stronger, absorbs shock well and is surprisingly flexible. The head is of hardwood, variable weight and tapered at one end. The length of the shaft depends on the height of the pony and the length of the rider's arm but is approx. 52″ long. Graphite and other materials can also be used.

THE TACK ROOM

There are several tack rooms, the two major ones housed in a spacious barn. The rooms are large and immaculate with low storage lockers running round the edge, good for storage and for sitting on. The walls are lined with cedar panelling to deter insects. Notices are in both English and Spanish so that although the setting is very English, the Argentinian players will also feel at home here. Large wheeled chests are for the team's equipment when travelling to matches such as to the snow polo in St Moritz and international competitions.

Players often bring their own favourite saddles but the design is generally one with a reinforced tree, a straight or forward front flap and no knee rolls. La Martina and Hermès are popular makers of polo saddles and use carbon fibre trees which reduce the weight. Stirrup leathers are of heavy-duty buffalo leather 1.25" wide (supplied by E. Jeffries, Walsall) which is virtually unbreakable and stirrup irons are generous to support the rider standing up in the stirrups during play. The saddle is secured with leather or synthetic girths, an elasticated overgirth

One of The Black Bears' tack rooms in Henley.

The bridles are mostly gag-snaffles with copper alloy bits. Various nosebands are used: drop, Grakle and cavesson as well as standing martingales. Draw reins which are fixed to the saddle at one end and run through the ring of the bit back to the player's hand can be used for extra braking power.

for safety and a breastplate to prevent it slipping back.

Ponies and riders are well protected. It is compulsory for ponies to wear bandages on all four legs during matches and most also wear tendon boots on the forelegs. Manes are hogged and tails braided to prevent them from catching in the reins or sticks. Riders must wear a helmet and usually have gloves and leather knee pads. Other safety features include quick-release designs of stirrup and padded, collapsible goalposts.

Most players in Britain play low- and medium-goal polo at one of the hundred or so polo clubs across Britain. A novice player would practise 'stick and balling' under tuition from a polo club which may also hire out ponies. The Pony Club junior teams compete at national and international level and the Armed Forces promote polo as a way of developing teamwork and leadership. The British Army Polo Association is establishing para-polo for rehabilitating young veterans and working closely with the Invictus Games inspired by Prince Harry.

Britain has many beautiful polo grounds, Windsor and Cowdray Park being two of the best known, where national and international tournaments are held. The season is from May to September and indoor or arena polo is played during the winter.

The Polo Game

Briefly, the rules are intended to safeguard the ponies and their riders.

There are two mounted umpires and a third standing on the sidelines.

The object is to hit the ball between the goalposts of the opposing team. Each team has four players. Men and women may play in the same team.

Any dangerous riding is deemed a foul, the penalty for which is a shot at goal by the opposing team from a distance determined by the severity of the foul. Yellow cards are given for disciplinary breaches.

The main rule is 'the line of the ball' which determines the 'right of way'. Players following the ball in its direction of travel have right of way. An opposing player may not cross in front but can attempt to push the first rider off that line by leaning his pony's shoulder against the first pony. The angle at which the ponies engage with each other, usually at great speed, is critical: an acute angle is safest, a broadside impact highly dangerous.

The game is divided up into chukkas, each of seven minutes duration with thirty seconds of

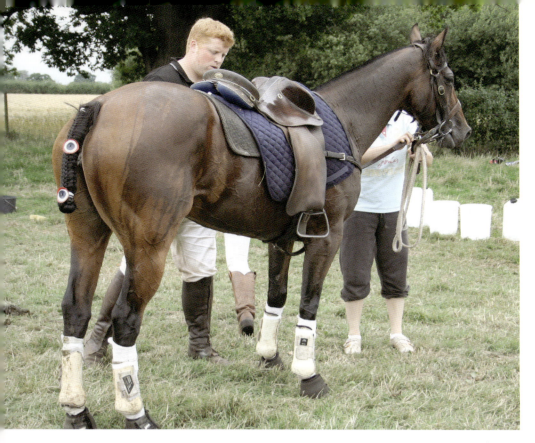

THE TACK ROOM

of white and made of leather or synthetic material which is inflated to the size of a mini football.

At the end of play there are usually three awards: for the most Valuable Player, for the Best Playing Pony and for its groom.

Good polo is a thrilling game for players and spectators alike. It depends on a surprising number of dressage manoeuvres, excellent teamwork, a high degree of balance in both horse and rider – and plenty of courage!

A polocrosse pony after being washed down following play. The Australian stock saddle has a broad fender instead of regular stirrup leathers to prevent entanglement and the saddle has a strong thigh support.

overtime play. A three minute break is allowed between chukkas to change ponies and five minutes at half-time. Ponies may also be changed mid-chukka. The number of chukkas, typically four to eight, depends on the level of polo.

Players may hook their opponent's stick either from behind or from the same side as the swing. The player must use his stick with his right hand only.

Each player's ability is rated by a 'goal handicap'. Beginners start with a handicap of -2 and work towards the maximum handicap of 10. In a tournament the total handicap score of the team members determines the level of polo.

In Britain, high-goal polo is played by a team totalling 17-20 goal handicaps, medium goal is typically 12-15 and low goal is a total team goal handicap of up to 8.

Polo balls were originally made from bamboo root but are now synthetic and about the size of a field hockey ball. For snow polo the ball is orange instead

Polocrosse

Polocrosse is best described as mounted lacrosse. The full game is divided into six chukkas (each is an eight minute period of play) and played with a sponge rubber ball on a grass pitch but, unlike polo, players can ride only one pony. The racquet is more

Polo is fun for all the family. A delicious picnic, a glass of Pimms and 'treading in the divots' between chukkas is a lovely way to spend the afternoon.

Polocrosse teams playing in a junior international tournament near Chester hosted by the Celyn Club.

like a long-handled tennis racquet than a proper triangular lacrosse stick. The idea is to keep the ball in the air as much as possible by passing and catching to avoid possession by the opposing team until it can be thrown or bounced (but not carried) between the goal-posts to score. Each team is made up of six players with three being on the pitch at any one time. Each of these three players has to stay within their area of the pitch (attack, midfield and defence).

Lacrosse itself originated among the native Canadians and Americans and was played on foot in a variety of forms, accompanied by ritual and ceremony. French missionaries described it in the 17th century as violent and un-Christian. It was modified by Europeans and by the time Queen Victoria was a spectator it was organised enough to gain her approval and to be incorporated into the curriculum of private schools in Britain.

The mounted form dates back to 1932 when it was promoted as encouraging riding skills and teamwork at the National School of Equitation at Kingston in Surrey. The sport was further developed in Australia to become the modern game and in Britain it is regulated by the UK Polocrosse Association (UKPA).

Polocrosse is a fast, exhilarating contact sport, graceful when the ball is in the air and extremely athletic when picking up the ball on the ground at a gallop. The ponies are taught to spin, stop and turn at speed and can be of any type or breed but should be well schooled and balanced. Because the stick is not just held downwards during play, as in polo, care must be taken that there is nothing in the pony's tack that may catch or hamper its use, such as hooks on curb bits, draw reins and bits with long shanks.

The Australian stock saddle, without a horn is widely used as it is robust and offers plenty of support, particularly for bending down to pick up the ball from the ground.

Teams compete at junior levels up to the Polocrosse World Cup for advanced adults. The Pony Club and local polocrosse clubs give tuition and schooling for the ponies to improve their agility. Many players enjoy 'social' polocrosse for fun and whatever the level of expertise it is a very good spectator sport too.

CHAPTER TWENTY-SIX

The Pony Club
(Patron: HRH The Princess Royal)

The Pony Club is made up of children and young people aged 3–25 years old and all shapes and sizes of ponies, so there is a very wide range of tack and tack rooms. Since there is no typical Pony Club tack room, I chose an exceptional one instead – one in inner London where a branch of the Pony Club and Riding for the Disabled (RDA) work together.

Wormwood Scrubs Pony Centre won the Queen's Award for Voluntary Service in 2015. This is an extraordinary achievement, though perhaps not so surprising as the Centre is run by its remarkable founder and director, Sister Mary Joy Langdon of the Sisters of the Infant Jesus order.

On a winter morning and still in darkness, Sister Mary Joy and her dogs walk a couple of miles across Wormwood Scrubs Common to the Pony Centre. Today, as for the last twenty five years, she will be instructing classes and organising events for the Pony Club and RDA from 7am until evening, six days a week.

In this combined teaching project, the Centre is a shining example of how the Pony Club and RDA organisations provide lessons not just for riding but also for life. As I watched the two groups of children going about their various tasks, it was clear that they gain enormous benefit from each other by being together in the same place.

The young people who come here do not have their own ponies and so learn on those at the Centre. Even in this urban environment, surrounded by Wormwood Scrubs Prison, the White City Stadium and Hammersmith Hospital, the highest standards can be achieved both in horsemanship and volunteering. In 2014 the junior Pony Club team from the Centre qualified for the National Championships in Horse and Pony Care and numerous individual awards and proficiency badges are gained every year.

The Centre, opened in 1989, has a full-size indoor riding school (built in 72 hours by television's *Challenge Anneka* team), stable yards with looseboxes, office, stores, tack room and two paddocks with tall shady trees. The indoor riding school allows lessons and events to be held from 8am until the evening all the year round and after working, the horses can be turned loose into the arena to enjoy a good roll in the sand.

A new Stable Yard Classroom within the riding school was opened by Clare Balding in 2015 where up to thirty children at a time can be introduced to riding and stable management through talks and demonstrations. A full-size, mechanical horse simulator electrically powered to trot and canter (CELT) provides a first introduction for children to the height and motion of a pony and a smaller wooden horse is a good static model on which to practise putting on a bridle.

At the far end of the room is a comfortable double stall where a pony or the Centre's two donkeys can be brought in for equine therapy, particularly helpful for autistic children.

Morning housekeeping at the Wormwood Scrubs Pony Centre. The Centre accommodates about twenty horses.

The tack room is a large converted Portakabin with three-inch thick, steel-lined walls. Inside it is tidy, clean and warm and smells just like a tack room should.

Two compartments are full of saddles and bridles and a third is a library with pictures round the walls and a table. There is a wide variety of saddles and bridles to suit the various sizes of pony and the capabilities of the children.

Special tack includes the gift of eight double bridles which bear the Loriners' crest on their browbands. These were a twenty-fifth anniversary present made by the Loriner's Company and are used by the Centre's Dancing White Horses, a mounted display team. Next to the padded vaulting roller with its large leather handles are about a dozen regular saddles which have been accumulated so that there is always one to fit a new pony. Mary Joy recalled an old saddler's advice: 'the saddle flap should flip up as easily as the page of a book' and all this equipment is beautifully maintained.

In addition, there is a full array of tack designed for the RDA here (described in the Riding for the Disabled chapter).

Daily briefings between Mary Joy, her deputy Bea Pike and their volunteer helpers are held in a large and lively reception room nearby with all-day tea and biscuits and in the afternoons the tack is cleaned here too with everyone helping.

Both the Saddlers' and the Loriners' Livery Companies fund Pony Club and RDA instructors' courses but a strong programme of fund-raising by the Centre members keeps the show on the road. Involvement with a film company (starring one of the ponies as a unicorn), dog shows, record-breaking attempts for the *Guinness Book of Records*, marathon running and displays by the Centre Team are just a few of their fund-raising events.

None of this would have been possible without the driving force of Sister Mary Joy, a practising nun and the kingpin of the Centre. Her vision, empathy and expertise touch all who meet her. Her order, Sisters of the Infant Jesus, was founded in 1660 in Rouen, France, to help street children find a productive life. However, it's not just the children of London who benefit from riding at the Centre but the young helpers too who find that working there gives them a leg up in life – they enjoy the camaraderie,

The Mechanical Horse (CELT) in the Stable Yard Classroom. Resh, a pupil trying out the horse for the first time assisted by Bea Pike WSPC trustee.

new responsibilities and the opportunity to take national educational certificates and Pony Club proficiency tests.

Mary Joy entered the novitiate for her religious order in 1984 and after becoming a junior professed sister, she worked at the London Lighthouse AIDS/HIV clinic.

In 1989 she was given the opportunity of starting a small riding school on the Wormwood Scrubs Park and around this time was approached by the Pony Club to start a new branch, now called the Wormwood Scrubs Pony Centre.

When Mary Joy first arrived, the site was a rough area of scrub with a muddy cycle track, a standpipe and concrete bollards. The first job she did was to dig out the bollards by hand and set about raising money. Not that she has ever been short of challenges – she was brought up on a farm and later joined the East Sussex Fire and Rescue Service as the first operational female firefighter in Britain where she stayed for seven years.

Early setbacks, such as the gypsies stealing her ponies, were overcome and now, twenty six years later, it is a haven for the 300 or so children who attend.

Two of Mary Joy's most recent accolades are that she was chosen to carry the Olympic torch on part of its relay journey round Britain in 2012 and in 2014 she was appointed an honorary member of the Worshipful Company of Loriners.

The Centre is affiliated to the British Horse Society which sets the national standards for instructors and Mary Joy is one of those responsible for training all the RDA instructors for London.

On my visit I meet a group of them who have arrived to learn by watching her give a riding lesson to four young special needs children. The ponies, their riders and helpers circle in one end of the large arena so that the children can easily hear Mary Joy and feel in contact with her. Her grey hair tidy under her riding hat, she moves with a youthful vigour encouraging the children through their exercises. Bambi, her dog, is an important part of lessons catching the tennis balls the children throw to improve their coordination. It is clear to all of us that Mary Joy's calm natural authority develops such a rapport that for each of the children every lesson is a special occasion and equally that every child is special to her.

The message here today is that when learning is fun and imaginatively taught, everyone feels a sense of achievement. The children not only learn to ride but with the pony as a catalyst, they develop balance, communication and confidence.

The horses and ponies here are of mixed types and sizes ranging from a Shetland through to larger ponies, cobs and Arabs, all perfectly behaved.

A beautiful 29-year-old pony, Sioux, rescued from Southall Market as a youngster and still scarred from her abuse, amply repays her good fortune here by providing equine therapy for disturbed children. She has also been painted by Lucian Freud (*Skewbald Mare* at Chatsworth House) and has taken part in

A stripped-down saddle revealing the interior to show how it is assembled.

numerous public displays. Another special senior resident is Quizzy, bred by Mary Joy and now 39-years-old and resting on his laurels of having given over ten thousand rides to children during his long service. Since my visit both ponies have passed away peacefully.

The opportunities that children find here may be life-changing for some. Mary Joy and the Pony Club share the same philosophy: to provide opportunities by learning through fun for anyone under twenty-five who has an interest in horses and ponies.

The Pony Club is a major source of the seedcorn for the future and a starting point for many professional equestrians, including Olympic medal winners, such as Ian Stark, Carl Hester and Mary

Clare Balding (left) talking to Sister Mary Joy in the tack room about the Loriners' gift of eight double bridles hanging on brass horseshoe fixtures alongside.

Guardian of the tack room. No mice here for Mr Ginger!

early grounding can be useful for a variety of first job opportunities, such as work riders, grooms, student jockeys and many others.

In 2014 a new initiative was launched by the Pony Club: The Rider Talent Pathway, which identifies promising riders (aged twelve or over) throughout the country and supports them with extra training from top class coaches. This scheme is for those interested in show jumping, dressage or eventing who aspire to become future international riders.

Proficiency or mini-badges are awarded for a variety of skills as well as for riding, for example, stable management and wildlife identification for younger children (aged 6–9) and volunteering and map-reading for older ones.

A plaque to celebrate the 25th anniversary of the Wormwood Scrubs Pony Centre. The Loriners' Livery Company (founded in 1261) is one of the most ancient in the City of London. Its members make the metallic components of saddlery and harness, such as bits, buckles and stirrups. The Loriners help to sponsor the Wormwood Scrubs Pony Centre and organise courses to teach the correct fitting and choice of bits.

King, none of whom came from riding families or owned ponies when they began. At the 2012 London Olympics, seventeen out of eighteen Team GB equestrian Olympians won medals and all but three of them started their training through the Pony Club.

The Pony Club (started in 1929) is now the largest youth equestrian organisation in the world and has an international programme of team competitions and exchanges across twenty-seven countries. Young people are taught to ride and to care for horses and ponies whether or not they own one. Riding can be enjoyed at many different levels with the option of joining in with nine different disciplines: tetrathlon, eventing, polocrosse, polo, dressage, mounted games, endurance riding, pony racing and show jumping.

In 1957, Prince Philip introduced the idea of mounted games or gymkhana events to encourage teamwork and to give children a chance to compete regardless of what kind of pony they had. The national finals are held at the Horse of the Year Show every year and spectators raise the roof with excitement.

There are national Pony Club tests for youngsters up to age twenty-five and courses for those who wish to go on to become career coaches. The Pony Prep (and Pony Prep Plus) online programmes are designed by the Pony Club to help members prepare for the proficiency tests at home and this

The Pony Club is also an important social network for children throughout the country especially in the rural areas and quizzes, musical rides, gymkhanas and Pony Club Camp provide a lot of fun as well as instruction. It is largely run by volunteers who receive training and support whereas coaching is done by accredited coaches. In 2017, the annual cost per child was £72 for Branch members and £29 for Centre members. Generally, Branches are for members with their own ponies or horses and Centre members use those provided on site.

Most importantly, many life skills are learnt: team spirit, sportsmanship, manners, self-discipline and confidence, pride in having high standards, communication with adults and respect for animals and the environment. For all children the opportunity to learn to ride in such a safe and supportive environment can be of great benefit and, who knows, may lead on to international riding careers, or to related professions such as farriery, veterinary medicine, equine podiatry and chiropractic, nutrition, saddlery making and instructing.

Alex, aged thirteen, a Pony Club member at WSPC, shows off some of her proficiency badges.

Flint and Denbigh Pony Club Branch at Mini-Camp (10 years and under).
Junior members preparing to have a tug of war with Dolly, a forestry horse hitched to her logging arch.
Dolly won and the children were bowled over by her huge horsepower!

CHAPTER TWENTY-SEVEN

National Hunt and Flat Racing

The world of racing is rich and vibrant, its long history so interwoven with jubilation and disaster, skulduggery and courage that it could fill a library. For the betting man there is hope of instant salvation: rags to riches from that lucky 100-1 outsider, though more fortunes have been lost than won.

Racing seems a natural combination of both the competitive streak in man and the swiftness of the horse. In the pursuit of speed and lightness (in contrast to the massive weights of over two hundredweight of armoured jousting) a race saddle may weigh only one pound – with so little between horse and jockey that they appear together more like a centaur.

The power of racing to thrill, whether on English turf, in the dust of the Australian outback or on the Mongolian steppes, never diminishes. Its colourful strands of high emotion, the hope of winning and the athletic beauty of the horses, can be an addictive experience.

Horse racing in Britain offers plenty of variety and every racecourse has its own character: the sophistication of Goodwood, the friendliness of Great Yarmouth, the charm of Cartmel, as well as the fun of local point-to-points with the spectators scarfed and hip-flasked against the weather. The Roman city of Chester claims to have held the first officially recorded race in 1539 and has the oldest racecourse still in use.

The elegance and royal patronage of summer flat racing on the turf gives way to the late autumn and winter ruggedness of steeplechasing. Royal Ascot is the most prestigious race meeting and attracts an international entry. The Cheltenham Spring Festival tests the rivalry between the top British and Irish horses over jumps and hurdles amid the party atmosphere of an Irish ceilidh. The Grand National, first run in 1839, is the most famous race in the world, watched by millions who may never have met a horse but are touched by the courage of runners and riders and the whole fascinating spectacle.

Many racehorses are celebrities who catch the imagination far beyond the racing world – Black Caviar, the outstanding Australian filly, appeared on the front cover of Vogue magazine and Eclipse had a full funeral at which cakes and ale were served and poems read. Shergar, who produced the most spectacular Derby win in history, was tragically kidnapped and presumed dead after his first season at stud and Man O'War was embalmed and thousands filed past his coffin.

Newmarket

The Jockey Club in Newmarket was founded in the 1750s and is the birthplace and global centre of modern racing. It was the headquarters for all racing in Britain until 2006 when regulatory responsibility was transferred to what is now the British Horseracing Authority. The elegant Jockey Club Rooms are filled with fine paintings, sculptures and racing artefacts, including the infamous blackball box which was used when voting for new members. In the forecourt is John Skeaping's bronze of Hyperion, the most successful British-bred sire of the 20th century.

Gibson's of Newmarket saddlers' workshop. In the foreground is a racing saddle and several reins. Paper patterns for cutting out leather for saddles hang on the wall.

James I built a hunting lodge for coursing greyhounds on Newmarket Heath and, for the amusement of his guests at his annual hunting party, he released a hundred hares and a hundred partridges. Gradually, races between the horses of his followers became as important as the coursing and the chalk downland of Newmarket Heath provided excellent going.

Charles I enjoyed racing here before Cromwell ploughed up part of it to mark his displeasure with the sport. It has since been so jealously guarded from attack by property developers, war, agriculture and transport links that the new railway was forced to tunnel underneath it. Charles II firmly re-established racing in Newmarket in 1666 by winning his inaugural Town Plate race which is an annual flat race of three miles six furlongs, ridden by amateur jockeys and is thought to be the world's oldest surviving horse race.

With about 3,000 horses in training in and around Newmarket (population about 15,000), and an estimate of one job in three related to the racing industry, a substantial support infrastructure is required. Newmarket has two of Europe's most advanced equine hospitals (Rossdale's and the National Equine Hospital) as well as one of the British Racing Schools, the National Horse Racing Museum, the Injured Jockeys' Fund headquarters and Gibson's the saddlers.

Besides the Jockey Club Rooms, another shrine to racing is Tattersalls who celebrated their 250th anniversary in 2016 as a premier auction house. Tattersalls' Newmarket bloodstock auctions are for flat racers but those in Cheltenham, Ascot and in Ireland sell chasers as well.

In their high-domed Newmarket sale ring with its steeply tiered seating, everything is immaculate: the circular walk of the arena brushed, every horse polished to a satin sheen. There is barely even the scent of horses, only the smell of money and the ozone of the racing world.

The sale of a thoroughbred is far from simple; these animals are not just commodities to be bought and sold from internet lists. The reason they are here in all their living, breathing, athletic beauty is so that their potential can be judged on the hoof. For those who can detect that flair in a youngster, see in its eye the confidence and the 'I am the best' attitude of a winner, there are fortunes to be made. It is not only the pedigree of a horse but its individual quality and presence that counts and experts called 'pinhookers' specialise in picking out potential winners from young stock.

Anyone may come to the sales but newcomer beware; sit still or you may go home with a very expensive racehorse. This is serious theatre, discreet and élite – no numbered paddles to hold up here. Around the floor are several smartly dressed young men their eyes fixed on the buyers. These 'spotters' are well briefed as to who is sitting where and can pick up the twitch of a catalogue or the scratching of an ear as a bid. The green LED screen high on the wall rolls over at 5,000 guineas at a time, orchestrated by the auctioneer's unique incantation, the tension rising tightly round his hovering gavel. In the December 2017 sales on one day the turnover was 45 million guineas.

Of the 5,000 foals produced for flat racing in Britain each year, less than two thirds will reach the start line of a race and of these, less than 10% will win. Perhaps fewer than half a dozen will become national heroes and household names.

Racing Saddlery

The Newmarket saddlery business was founded by Colonel Gibson in the early 1900s and received its Royal Warrant in 1932 as the exclusive suppliers of racing colours to HM The Queen and the Royal Family. Gibson's tack shop caters for all types of riders and their workshops produce a wide range of racing and endurance saddlery for both the retail market and special commissions. Besides the traditional handcrafted saddle making, they can produce bespoke pink leather saddles and white bridles decorated with Swarovski crystals to extend the tack range.

Behind the scenes of EJ Wicks Saddlers (est. 1902) in Lambourn. Master Saddler Tiffany Parkinson (left) is demonstrating the Australian 'Stride Free' exercise saddle (centre). Placed on top of the saddle is an example of its inner white moulded plastic tree. As the saddle warms up during exercise, the plastic tree inside becomes more flexible. Saddler Erin Pope (right) is repairing the rubber grip on a pair of racing reins.

The development of lighter, stronger saddle trees is ongoing. Saddles have become tiny compared with the full size ones shown in paintings up to the 1850s which were for jockeys who rode with long stirrup leathers and an almost straight leg. As steeplechasing became more popular, stirrup leathers were shortened and jockeys rode in a crouching position ready to conform to the horse's shape if jumping, relying on balance rather than the saddle to stay aboard.

Racing tack needs to be lightweight for flat racing and more robust for National Hunt racing. Two types of saddles are used: one for exercise and one for racing. The exercise saddle has a half tree and a big enough seat for the jockey to sit down on. A popular model here, produced in Australia, is the 'Stride Free' in which the metal half tree is replaced by a moulded plastic form. This is lightweight, distributes the weight away from where the points of a metal tree might press on the shoulders and allows unrestrained movement. The arch or head over the withers is reinforced with steel to maintain the shape of the saddle.

Saddles for flat racing weigh from eight ounces to three pounds while jump saddles weigh about six pounds. Their tiny size allows the horse's shoulders and back to move to their full extent. Flat race saddles have a half or full tree of plastic and no stirrup bars, the stirrup leather being looped over the side bar of the tree. There is light felt padding and a saddle cloth is placed underneath.

Pockets for lead weights can be added to the tree. The jockey barely sits on the saddle, his weight being taken by the stirrups and distributed either side of the horse's spine by the rigid tree. An elasticated girth and overgirth, together with a breast strap, make sure it is secure. Stirrup leathers can be made of rawhide to prevent stretching and reins are usually stitched on rather than billeted (steel hook) for safety.

Bridles are minimal though may have sheepskin nosebands and cheek pieces attached to help the horse lower its head or keep a forward focus. Flat racing bits tend to be simple jointed snaffles but for jumping, where more control is needed, the effectiveness of the bit and noseband must match the character of the horse. A chifney bit (designed by the jockey Samuel Chifney who won the Derby in 1789) with a thin metal ring attached to the centre of a snaffle bit and circled beneath the lower jaw, can be used to give more control over fences.

Soft hoods, some with partial blinkers, may also cover the ears and can be worn as a calming device.

Reins have various designs, usually covered with a textured rubber sleeve, to make them easy to grip despite sweat or rain, and are longer in length for steeplechasing than for flat racing. A bib martingale (a small sheet of leather joining the reins under the horse's neck) is used to keep the reins together if the jockey should fall off during the race.

Scrupulous hygiene and biosecurity is vital to avoid infection where so many horses are stabled in close proximity and this applies particularly to the care of all tack. Bits are thoroughly scrubbed and girths may have fabric sleeves which can be removed and easily washed.

Flat racing

Briefly, flat races are divided into two main categories, Handicaps and Conditions races (Group 1, 2 and 3 and Listed, allocated by ability, Group 1 being the best). Handicap races are divided into classes also on the basis of ability but each horse is

Returning from the gallops at Park House, Kingsclere, Hampshire.

given a weight to carry to make the race fairer and more competitive. The British Horseracing Authority (BHA) Handicapper bases the weights (reconsidered for each race) on a number of criteria such as distance, the state of the ground, the calibre of the other runners and past performance. From these, a rating is given to every horse so that one may be compared with another, a measurement particularly appreciated by the betting public.

The Derby, held at Epsom over twelve furlongs, is the richest and the most prestigious of the five English Classic flat races for three-year-old, top quality (Group 1) thoroughbreds. The others are the 1,000 Guineas, 2,000 Guineas, St. Leger and the Oaks. In 1779 Lord Derby won the toss of a coin with Sir Charles Bunbury to decide which of them should give his name to this new race. The Oaks, first run in 1778 for fillies, is named after one of the 12th Earl of Derby's estates. These races are the highlights in the social calendar of flat racing enthusiasts just as the Cheltenham Festival and Grand National meetings are for National Hunt racing followers.

One of the most famous flat race jockeys is Lester Piggott who first raced as a ten-year-old and retired aged sixty having won 4,493 races, including nine Derbys. Another is Lanfranco 'Frankie' Dettori, known for his flying dismount in the winner's enclosure, who arrived at Luca Cumani's yard in Newmarket from Italy aged fourteen. His talent and flamboyance have been good for racing and his seven wins in seven races on a day in 1996 at the Royal Ascot Festival remain unrivalled. His son, Rocco, having made a good start in pony racing may well become a third generation Dettori jockey.

Andrew Balding, Trainer at Kingsclere, Hampshire

As a contrast to the intense 'hot house' of Newmarket's horses, people and traffic, the historic Park House Stables in Hampshire are surrounded by beautiful open countryside and the hundred-year-old turf on Watership Down.

British and international flat race champion trainer Ian Balding took out a licence here in 1964 and sent out two thousand winners on the flat and over jumps before handing over to Andrew, his son and assistant trainer, in 2003. Andrew Balding has expanded the yard and in 2017 trains about

A corner of the colour room, a former chapel, at Park House Stables. The portrait above the Victorian jockeys' weighing chair is of Glint of Gold by R.Driscoll.

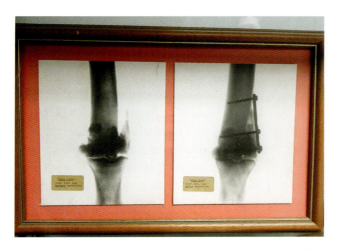

An X-ray of Mill Reef's fractured foreleg before (left) and after surgery (right) which put an end to an outstanding flat racing career but from which he recovered to prove himself an exceptional sire.

170 horses looked after by seventy staff. Winners here now consistently bring in more than a million pounds of prize money a year racing in the UK and around the world.

The facilities are superb; tree-lined avenues connect the yards to the gallops and there is no traffic. There are turn-out paddocks, a swimming pool and treadmill, open and covered lunging areas and a one furlong covered indoor ride. The seven different yards are named after winning horses, making a total of 175 looseboxes, ranging in style from the original Victorian brick to modern American barns. Each yard has its own tack room.

The main tack room is also called the colour room because of a brilliant array of jockeys' racing silks taking up a whole wall. The room is a former Catholic chapel built for the use of the many Irish lads who worked and rode here in the 19th century. Now stripped of pews there is plenty of room for a large amount of tack, including the sheepskin nosebands worn by Kingsclere horses, and unusual period pieces from Victorian times, such as a jockeys' weighing chair.

All the tack needed to accompany the horses to the racecourse is brought here from the various yards and checked. A large chart shows which colours or silks belong to each of the current owners and these are put into special leather bags designed by Ian Balding, ready to travel with the horse on the day.

The nave has also been used as a gym but its most important use was as an operating theatre for Mill Reef, owned by the American millionaire Paul Mellon, and trained by Ian Balding.

Mill Reef (1968–86) was an exceptional horse, winning twelve of his fourteen races and the first to win the Derby, King George VI and Prix de l'Arc de Triomphe. On a routine training gallop as a four year old in 1972 he stumbled and shattered his foreleg, sustaining multiple displaced fractures. He was moved with difficulty to the colour room where an improvised operating table of straw bales covered in polythene had been quickly assembled in the centre of the room. The vet, Professor Jim Roberts, told Ian that they had only one chance to anaesthetise the horse so that he fell correctly onto the straw bales. If he missed, he would have to be put down. All went well, the cannon bone was pinned and braced and Mill Reef recovered sufficiently to retire to the National Stud where he sired numerous successful offspring including two Derby winners.

Thoroughbreds

For flat racing, horses must be thoroughbred, DNA sampled and registered with Weatherbys, keepers of the General Stud Book for British and Irish thoroughbreds since 1791. Weatherbys' Stud Book preceded the first edition of Burke's Peerage by 35 years, equine pedigrees trumping the record of human aristocracy.

The three foundation stallions, imported in the 17th and early 18th century, to which, by common acceptance, all thoroughbreds can trace their lineage are the Byerley Turk, the Godolphin Arabian and the Darley Arabian.

Artificial insemination is not permitted in thoroughbred breeding and the top stallions travel the globe covering mares in the spring in the northern hemisphere and in the southern hemisphere in the autumn. A thoroughbred's official birthday is 1st January in the northern hemisphere and the 1st August in the southern hemisphere.

Thoroughbreds can be very talented and are often highly strung. It is the trainer's job to understand the psyche of each individual horse in his care, as well as to develop its physical power as an athlete. It is considered to be an art as well as a science and trainers with an intuitive skill include the late Sir Henry Cecil of Newmarket and Aidan O'Brien, trainer to the Coolmore Stud Associates in Ireland who head a list of hundreds more. Sir Henry used to wait until his yard was quiet and deserted in the evening and then would observe every horse in its loosebox. Its demeanor and position in the box informed his management of each one.

Thoroughbred owners and breeders range from royalty and millionaires to syndicates and small individual breeders. HM The Queen takes an active interest in racing and is an expert owner and breeder. Her Sandringham stud is one of the oldest racing studs in the world.

The magic and mystery of breeding has not been entirely unravelled by science – yet. A recent genetic test claims to categorise two-year-olds into their optimal race distances. The test is based on the levels of a protein, myostatin, which appears to influence the mass and predominant type of muscle (fast twitch fibres) during growth. It may be that by tailoring the training of young horses to their predicted best performance (sprint, staying or middle distance), fewer injuries and greater success can be achieved.

It is still possible for the small breeder to hit the jackpot but they would be a David to the Goliath of the big breeding studs, such as Coolmore (Ireland), Juddmonte Farms and Darley (Newmarket), with their might and millions. One successful small breeder was the subject of a film *The Dark Horse* in 2014. A publican and her husband, Jan and Brian Vokes, bred a colt (Dream Alliance)

The bronze centrepiece of Mill Reef at Park House Stables by John Skeaping, one of many memorial buildings and sculptures of him in the UK and USA.

Frankel, ridden by Tom Queally, winning the Qipco Champion Stakes at Ascot in 2012 over 10f. This was his last race before retiring to stud.

from a £300 mare and reared it on their allotment. They raised a syndicate from 22 local friends who each paid ten pounds per week to cover expenses for training Dream Alliance as a steeplechaser. The horse confounded the critics and had a lucrative career which culminated in winning the Welsh Grand National. It was also the first horse to win a race after undergoing stem cell treatment for a damaged tendon.

In the 1970s Robert Sangster and John Magnier duelled with the ruling families of Saudi, Dubai and Qatar across the sale rings of Keeneland, Kentucky and Saratoga, NY, sending the price of yearlings to over $2 million dollars apiece. Their strategy was to find a yearling which would become a pre-eminent stallion by winning the Classic races. The thoroughbred world would beat a path to its door and through shares and covering fees such a horse would repay its owners many times over.

By the mid 1980s this mad bravura had ended and mass bankruptcies followed. The market has recovered and in 2017 the sprinting filly, Marsha, was the target of a duel between Godolphin and Coolmore with the hammer falling on a bid of six million guineas from John Magnier for Coolmore.

Winning mares and stallions command substantial fortunes. Frankel (by Galileo out of Kind) is considered to be one of the greatest racehorses ever. He is owned and bred by Prince Khalid Abdullah and was trained by the late Sir Henry Cecil in Newmarket. Known for his big feet, large appetite and huge stride of twenty four feet (twenty one or two feet is normal), he retired to stud unbeaten as a four-year-old with a record rating of 147. His stud fee began at £125,000 and in his first season his mares achieved a 95% conception rate, with his first filly foal selling for £1.15 million.

Such exceptional racehorses are rare and it is interesting that three historically fast horses, Eclipse (British), Pharlap (Australian) and Secretariat (USA) all had very large hearts at autopsy. The average weight of a horse's heart is three and a half to four kilograms (about nine pounds) but Eclipse and Pharlap each had hearts weighing over six kilograms – about 25 percent larger than average. Secretariat's heart was estimated to be almost twice the average size by Dr Tom Swerczek who performed the autopsy.

A winning stallion is no guarantee of a successful sire. El Gran Senor, after a dazzling career, had low fertility as did Al Kazeem who was returned

The Grand National. AP McCoy (left) on Shutthefrontdoor jumps The Chair.

to the racetrack. And mares can be unpredictable: Impression was a mare so slow she couldn't get out of her own way yet produced foal after foal who won Classic races – this type of mare is known by breeders in America as 'a blue hen'. Secretariat was the fastest mile and a half horse ever and winner of the American Triple Crown (Kentucky Derby, the Preakness Stakes and the Belmont Stakes for three-year-olds). In 1973 at Belmont he broke the world record by completing a mile and a half in 2.24 minutes and finished so strongly by thirty-one lengths that he needed at least a furlong to pull up. It is said that 5,617 winning Tote tickets were not cashed in on that day but were kept as souvenirs to mark such sporting perfection. He became a good but not great sire; about 9% of his 650 foals were stakes winners.

As a highly-tuned racing machine, the successful thoroughbred must combine an excellent cardiovascular system with a formidably strong musculoskeletal system.

Studies have shown that the racing of two-year-olds is not as injurious as was once thought. It is now accepted (even by the RSPCA) that careful training can beneficially model the developing bone, with the result that horses who were trained as two-year-olds were actually less vulnerable to injury in their later careers than horses who were not. Too much work too early however, can lead to stress fractures which are one of many factors responsible for the high failure rate. The Racehorse Owners Association report (2012) estimates that of the 24,000 horses in training that year, only 17,500 actually made it to the racecourse.

The Kurtsystem at Kingwood Stud in Lambourn has a technological approach to developing young racehorses. Each horse is harnessed into a machine, which is suspended from a monorail above the gallops, and its heart rate, respiration and blood pressure are monitored during training. The results influence the management of each individual, tailoring the workload to their capability.

The thoroughbred is bred to race. It has the bone quality for strength, a higher proportion of muscle to bone than other breeds of horse and a highly efficient vascular system for oxygen delivery. The muscles need enormous amounts of oxygen ranging from about 70 litres/min. at rest, to 1,000 litres/min. at the gallop. Beneath the resplendent, lather-splattered coat of the galloping thoroughbred lies a monstrous heart trilling at four beats per second, each minute driving over four hundred litres of oxygen-rich blood from lungs to muscles.

With the exception of rare individuals with larger than average hearts, the evolution of the thoroughbred's capacity for speed has probably stabilised due to its relatively small genetic pool. Course records are not likely to be broken often.

The first of the thoroughbred progenitors to arrive in 1687 was the Byerley Turk thought to have been captured during the Siege of Budapest in the Balkans. He takes his name from his last owner, Colonel Robert Byerley, who campaigned him at the Battle of the Boyne before retiring him to stud at his Goldsborough estate in Yorkshire. His sire line has dwindled but continues today through the late Compton Place and Dunaden, winner of the Melbourne Cup and five million pounds in prize money.

The Darley Arabian was imported in 1704 by Thomas Darley, British Consul in Aleppo, Syria. His influence has spread far and wide with many winners, such as Eclipse and Whistlejacket, famously painted by George Stubbs.

The Godolphin Arabian is thought to have been foaled in the Yemen, exported from Syria to Tunis and presented as a diplomatic gift to Louis XV of France. In 1730 he was sold on to Mr Edward Coke in Derbyshire and eventually to Francis, the second Earl of Godolphin. His grave is believed to be in what was the Godolphin Estate, now Wandlebury Country Park, near Newmarket where a large slab inscribed *'The Godolphin Arabian, Died in 1753. Aged 29'* lies in the shelter of an archway. It is said that the stable cat with which the Godolphin Arabian had a strong friendship, is buried with him. Although no records of his racing exist, the quality of his progeny such as Man o' War was soon evident.

National Hunt Racing

National Hunt racing or steeplechasing, which runs from November to April, has a special place in the hearts of the British racing public. The Cheltenham Gold Cup is the holy grail of jump racing and winners include Arkle, Kauto Star, Desert Orchid and Best Mate who inspired a mass following. Arkle was known as 'Himself', in Ireland and was a national hero with 'Arkle for President' painted on city walls.

It is a tough sport and both jockeys and horses command great respect for their courage and skill. Sir Anthony McCoy has an unbeatable record of being champion jockey for sixteen consecutive years but has also survived scores of injuries. Jumping at speed is hazardous. Jack Berry, ex-jockey and founder of the rehabilitation and training centre, Jack Berry House in Malton for jockeys, has pointed out that jump jockeys average a fall every sixteenth ride.

It takes several years longer, with added cost and risk, to produce a horse for hurdling or 'chasing

Nicky Henderson's main yard at Seven Barrows, Lambourn during his annual Open Day. The most recent winners are paraded, interviews given to the national press and the general public welcomed.

than it does for a horse racing on the flat. At the age of six or seven they are considered strong enough to jump safely while a horse racing on the flat may do so at two years old.

The Grand National is a unique race, probably the greatest equine challenge in the world. Some forty qualified horses and jockeys compete over four and a half miles and thirty jumps for a prize fund of about a million pounds. Held at Aintree in Liverpool there is certainly no English reserve in the roar of the crowd when the flag drops and runners set off like a cavalry charge for the first fence. The tension is electric, everyone enthralled.

Grand National history is packed with extraordinary events, people and horses. Probably the most exceptional horse was Red Rum, a British legend who is buried near the winning post. He won three times in 1973, 1974 and 1977 and was second in 1975 and 1976, a magnificent achievement for a horse with problem feet. He was locally trained by Ginger McCain and exercised on Southport beach. At Christmas time in 1977, a crumpled page from a child's school exercise book was seen stuck to the waiting room wall at Liverpool's Lime Street station. Scrawled in crayon was written, 'Who was born in a stable and known to millions?' On the line below was added, 'RED RUM!'

Foinavon's win in 1967 at odds of 100-1 was extraordinary but perhaps one of the most popular and inspirational wins ever in racing history was by Bob Champion riding Aldaniti in the 1981 Grand National. Two years earlier, Champion had been diagnosed with cancer and given only months to live but recovered with chemotherapy despite complications. Aldaniti had had serious leg injuries requiring months of box rest, yet together they came home by over four lengths to a crowd wild with emotion. The pair went on to raise over fifteen million pounds for cancer research, Aldaniti living in busy retirement until he was twenty-seven.

Nicky Henderson, Seven Barrows, Lambourn, Berkshire

Nicky Henderson has been National Hunt champion trainer four times. His horses are almost all steeplechasers and hurdlers and have the luxury of training on the wide sweeps of chalky downland under huge skies. Henderson escaped from the prospect of either an army career or a life of stockbroking and began training in 1978. Through working with Fred Winter he met Corky Browne who became his head lad and together they have sent out a stellar cast of winners. Since moving to Seven Barrows in 1992, many of the horses trained here have lit up the annual Cheltenham Festival and become household names: Bob's Worth, Sprinter Sacre, Long Run, Altior, My Tent or Yours, and many more.

On the way to the tack room we pass a large board on which all the names of the horses currently in training are written. 'This', explains Henderson, 'is the pivot of the whole yard. This is where I match each horse with its exercise rider and determine its training regimen for the day. It's our central reference point for all the work.' With about 150 horses in training here, the theoretical permutations of horse and rider run into thousands of different combinations.

In the yard are three traditional tack rooms: two for exercise tack and the racing tack room for everything needed on race days. Like most tack rooms, they can evoke many equestrian and human personalities.

Two particular bridles in this racing tack room immediately conjure up the horses which have left their mark in history. They have given their names to the popular 'Grakle' noseband and the much rarer 'Citation' bit. Grakle won the Grand National in 1931 and the 'X' shaped strap arrangement of the noseband, buckling above and below the bit, was designed to stop him from crossing his jaws and evading the action of the snaffle bit. It is a noseband now very widely used in many equestrian disciplines.

Citation was an American racehorse and winner of the 1948 Triple Crown. He was almost unstoppable and the double bit, the thinner of which is attached directly to the noseband, specially

Nicky Henderson in his racing tack room. Behind him, his daughter Sarah rides out on Hunt Ball. Only race day tack and equipment is in this room. The bridles have a variety of bits and their colour-matched browbands will travel with each owner's silks in a leather bag on race day. There are no saddles in this tack room: professional racing valets look after the jockeys' personal saddles and make sure all the correct colours, logos and equipment for each horse accompany the jockey to the racecourse.

designed for him gives such leverage that, in Corky's words, 'It would stop a train'. It is only used on the most persistent pullers for their own safety and in expert hands. It was worn to good effect for a short period by My Tent or Yours (and also by Ian Stark's irrepressible Murphy Himself at the Badminton Horse Trials).

As we talked in the tack room about Henderson's earlier life and family army connections, he recalled a third historic horse which his godfather, Field Marshall Montgomery, was pleased to sit on for a propaganda photo opportunity; this was the favourite white Arab stallion captured from Erwin Rommel at the end of World War II.

The stallion caught the eye of a brigadier who asked if he could ride him. Nicky Henderson's father, Johnny, who was aide-de-camp to Monty at the time, agreed. However, the brigadier was not popular with the liaison officers and Johnny Henderson arranged for the brigadier to go out hacking in the company of another officer riding a mare who was in season.

When Nature inevitably took its course, the brigadier jumped for safety and ended up entangled in the branches of a tree while the stallion set off in hot pursuit of the mare.

More rare and ingenious bits were shown to me, such as a double bit linked together, its name unknown, which was designed to prevent a horse from hanging to the left or right during a race. Another was described as a narrow tube of a bit much wider than normal with an internal sliding bar which successfully steered Lady Lloyd Webber's Raymylette to several victories. The final choice of bit for each horse is made as a result of trial and error during training and with feedback from the exercise riders and jockeys.

This collection of unusual bits goes to show what lengths a good trainer will go to in order to find exactly the right bit for each individual runner – just one of many ways in which the trainer translates hope into success for his owners.

Research and Welfare

Veterinary research into equine physiology continues to inform the management of the racehorse, such as in the avoidance of gastric ulcers and the reduction of bone problems. Monitoring with wireless telemetry, which allows the visualisation of the respiratory system and the measurement of heart rate during galloping, is a major advance for training and more turn-out time in paddocks is thought to be beneficial to equine well-being. From April 2017, 10% of every bet placed goes to help fund the racing industry and related research, together with contributions from the major studs and various racing associations.

Standards of welfare have greatly improved for horses when they retire from racing and a number of organisations help to re-home them. The Thoroughbred Rehabilitation Centre in Lancashire and the Greatwood Centre in Wiltshire were the first to begin retraining ex-racehorses for alternative work, such as dressage, polo or eventing. At Greatwood these horses may also participate in equine therapy programmes for children and adults with special needs (*see* page 193). The Retraining of Racehorses organisation is the BHA's official charity and helps to support many of these initiatives.

The British Schools of Racing (in Newmarket and Doncaster) have comprehensive training programmes which also advise on welfare for jockeys and those working in the industry. Research into racing-related issues such as diet and the effects of concussion are ongoing and two excellent rehabilitation centres run by the Injured Jockeys Fund (IJF), Oaksey House in Lambourn and Jack Berry House in Malton, will be followed by a third in Newmarket, named after Sir Peter O'Sullevan, the distinguished commentator. HRH the Princess Royal is the Patron of the IJF and she and Prince Charles both raced competitively.

Racing is the most available of all equestrian sports to the public who can take part as owners, riders, punters and spectators. This billion pound industry is based not just on expertise at every level but also on hope – British racing is not only the sport of kings but also the sport of dreamers.

Returning to Nicky Henderson's yard on an autumn morning in Lambourn. One of the skills of a trainer is to match the horse with the most suitable rider for training and continuity. The careful selection of which horses to exercise together, especially when doing a piece of work, can lift a horse's competitive edge.

Riding for the Disabled Association (RDA) and Therapy with Horses

The therapeutic value of horses has been recognised for a long time, well before Sir Winston Churchill's famous quote, 'There is something about the outside of a horse that is good for the inside of a man.'

Dame Agnes Hunt and Olive Sands were among the first to introduce riding to help rehabilitate First World War soldiers with physical and psychological injuries. The medical profession were slow to be convinced but took note when Lis Hartel, a Danish rider affected by polio, won an individual silver medal for dressage at the Helsinki Olympics in 1952.

The idea gathered strength, propelled by some remarkably forward-thinking individuals and in 1969 became the charity Riding for the Disabled (RDA) with HRH Princess Anne as Patron and Lavinia, Duchess of Norfolk as President.

In 1996 the RDA competed in the Atlanta Paralympics for the first time and it has continued to thrive. Today in Britain it is a federation of approximately 500 independent groups which support about 26,500 adults and children with additional needs each year. Trained and qualified coaches organise riding, carriage driving and vaulting with the assistance of some 18,000 volunteers who also do a good deal of fund-raising.

The RDA has spread across the world where the dedication and resourcefulness of the coaches is extraordinary despite the lack of facilities in some places.

Pam Jones, a Fellow of the RDA and Equine Manager at the Clwyd Special Riding Centre, has travelled abroad advising local organisers and their groups. She described a child in Kenya with cerebral palsy who had arrived for her first riding session, 'She was waiting for me under a tree with a smile that lit up the shade and was dressed in her best frilly party dress and ballet shoes in matching lemon yellow.' A few ex-racehorses were led out of the bush into a clearing and a row of sticks stuck into the ground. The little girl was lifted up onto the saddle cloth and held as the pony was led in and out of the sticks. The doctors and physiotherapists could see what the benefits would be to this girl from the movement of the horse and her absolute joy moved the bystanders to tears.

RDA groups in the UK vary: some have the use of an indoor school, others use the corner of a field dodging thistles and cowpats in summer and bringing hot drinks in thermos flasks in the winter, but the cheerful and positive atmosphere is common to all. We volunteers (three per rider, one leading the pony and one walking either side) always felt better in ourselves after an RDA session. The improvement in the children's mood and ability was uplifting and

Alan's smile brightens the day as Harvey pulls him and Zena, able-bodied whip, through the water splash at the Clwyd Special Riding Centre in Flintshire. Alan holds the brown reins, connected to Harvey's headcollar while Zena drives with black reins attached to the Liverpool bit.

we appreciated the camaraderie and the perspective it gave to our own lives. The motivation, progress and pleasure it gives to those receiving horse therapy is inestimable.

The Clwyd Special Riding Centre Ltd. (CSRC), Llanfynydd, Flintshire, North Wales

The CSRC is indeed a very special centre. Its traditional Welsh stone barns, modern indoor school and stabling barn are set in thirty-six acres of fields with adapted accommodation for the holiday groups. The atmosphere of warmth and fun is immediate in the large, bright common room with games tables at one end and a hatch through to a kitchen at the other. Several children and parents are there talking and laughing over tea and biscuits and outside the sun is shining on the surrounding Welsh hills.

On the walls are photographs of the 'workforce': these are all the ponies and horses on site, twenty-four of them who have been donated, purchased or loaned (including one from World Horse Welfare).

The Centre was the vision of Anne Sopwith and her friends who together bought the property, a dilapidated farm, in 1981 and established it at great financial risk to themselves. Support followed from Marks & Spencer, Laura Ashley and Airbus which donated a hydraulic mounting platform. HRH The Princess Royal officially opened the Centre in 1992 and the Trustees continue to fundraise for the £5,000 required per week for running costs.

This is one of the very few centres in the UK which has the facilities and expertise to support people with any disability and can offer residential accommodation for holiday groups of riders and helpers.

Every week, over two hundred children and

adults, most referred from schools or clinics, benefit from the freedom and stimulation of the activities here. Many of the young people progress to achieving national qualifications and to competing in regional and national events.

Nicola Tustain, Para-Olympic gold medal winner in Sydney in 2000 and Athens in 2004, three times National British Para-Dressage champion and winner of numerous awards, is an Alumnus of the CSRC. Nicola was born with a paralysed right side (hemiplegia) and suffers from painful spasms so rides without stirrups and with one hand. She learnt to ride here aged ten and by seventeen she was competing internationally. She is now a freelance instructor, trainer and motivational speaker. She and Albion Saddlemakers in Walsall (*see* page 8) worked together to develop a saddle which would give her the extra support she needed. Paul Belton, Director of Albion explained the three principles of any successful bespoke design which apply to both horse and rider, 'Comfort, so that there are no distractions, stability so that muscles are not tensed unnecessarily and freedom to move as required.' For Nicola's saddle, this was achieved with the use of lightly padded blocks and Velcro panels in the saddle flaps so that she felt secure but still had flexibility of movement. The reins were joined with a horizontal bar so could be held in one hand.

Riding, carriage driving and vaulting all improve balance, co-ordination and confidence. Being at the Centre encourages sociability and participation especially in team activities such as vaulting. Wheelchair users feel particularly empowered to be looking down from a pony instead of up to people and parents too can meet here and support each other.

An additional element of CSRC is CELT (Centre for Equine Learning and Therapy) with state-of-the-art facilities including a fully simulated mechanical horse, smart board and a variety of multi-sensory resources.

The facilities at the Centre are so good that able-bodied groups also come to train here, providing a useful income stream. Dressage riders in particular find the CELT mechanical horse (Grand Prix model) helpful for improving their riding position. This full size horse is computerised to respond to heel and hand aids and simulates the movements of walk, trot, canter and gallop. The horse's software is able to convey on a large screen the action that particular aids would produce in a real horse, for example, a half-pass. It is also possible to programme different scenery onto the screen with a choice of 'riding' on the beach, up a hill or in a wood where the horse can be guided around the trees.

The Centre is busy all the year round with local, national and overseas persons with additional needs, with or without their parents and carers, such as the groups from Kosovo, Russia and Hong Kong who have come here for a holiday. Local RDA groups also use the facilities at the Centre. Their motto is 'It's what you can do that counts' and aims to turn 'I wish' into 'I can'.

The Wrexham Driving Group provides coaches and helpers for regular carriage driving sessions at the Centre. The Group members commissioned the first purpose-built vehicle for RDA use from

Bunny Ear (top) and Loop (bottom) rein adaptations to encourage a mid-prone hand grip similar to holding a cup or a pencil.

Bennington, carriage makers who hold the Royal Warrant. The result was a four-wheeled 'Fun Bug' which is a great improvement on the two-wheeled gigs previously used which had to be re-balanced for each driver. It has a detachable ramp for wheelchair drivers and additional rear wheel steering gives greater manoeuvrability in the confines of an indoor school when the weather is bad.

Lynne Munro, Chartered Physiotherapist and Hippotherapist, a coach and examiner, works one day a week at CSRC. She describes how hippotherapy (physiotherapy using horses) can be helpful for patients with muscular and coordination problems, 'Sensory input and stimulation is provided by the horse's gait which moves its rider in three dimensions: side to side, up and down and rotationally. This mimics the movements used in normal walking with the pony providing extra stability: four legs instead of only two. To maintain balance the patient uses core muscle groups which can be strengthened when the horse moves in circles and with transitions in pace. Sitting facing backwards involves more reliance on the vestibular balancing system and simply lying down along the horse's back is pleasurable and stimulating for some. The choice of saddles, sheepskins or small rollers with leather handles is selected for each patient.'

The Chartered Physiotherapists in Therapeutic Riding and Hippotherapy (CPTRH), a national network, runs a post-graduate hippotherapy course annually at CSRC. There is also an international organisation, Horses in Education and Therapy International (HETI) which links the RDA and related institutions as well as promoting research.

Army veterans from the Falklands, Afghanistan and Iraq come here too. One 23-year-old soldier who has lost both legs above the knee from an IED (improvised explosive device) was very apprehensive about the idea of getting on a horse, even with the aid of a hydraulic platform or hoist. Eventually he agreed to try and found he thoroughly enjoyed being on the horse saying, 'I lost two legs and gained four.'

The design of tack for riders with additional needs is innovative and flexible, sometimes specially commissioned from a local saddler to accommodate those with restricted physical movement. Bethan, aged ten, has cerebral palsy. By gradually helping her to relax the rigid grip on a horizontal bar section of rein to a more normal handgrip with thumb uppermost, and to keep her upper body more erect, she will have more control over the pony. This of course has importance beyond riding: this is the coordination needed to pick up a cup and drink from it. To encourage this mid-prone hand position, Lynne Munro designed 'bunny ear reins'. These are leather, finger-like vertical grips, one for each hand which are attached to the D-rings either side of the pommel and can be held together with the rein. Loop attachments are helpful for riders whose fingers cannot grip easily.

Many children are more comfortable riding on a sheepskin (secured with a surcingle) rather than on a saddle as it allows better contact with the movement of the horse. Rainbow reins with bands of colour are easier to hold at the correct length according to the colour code indicated by the coach, than plain reins. They are usually clipped onto the pony's headcollar, worn under the bridle, rather than onto the bit. All the saddles have breastplates as well as girths to keep them securely in position. Toe stoppers (fixed plastic cages) are attached for safety to the stirrups to prevent the foot slipping forwards through the stirrup as many riders are unable to wear riding boots with a heel.

This is a place where huge contributions are made, not only by the staff and volunteers but by the riders themselves. They inspire one another, relax and make new friends, relieved of the competitive challenges in their day to day lives. The ponies too, especially if they are rescue cases and have been loaned from World Horse Welfare, find a haven here where they can lead a useful and happy life.

There are many examples of the beneficial effect horses have on autistic children such as promoting calm and encouraging non-verbal children to communicate. To see an eight-year-old boy whispering 'Thank you, pony' at the end of a session, knowing they were the first words he had spoken for many months, was profoundly moving. Traumatised children will often unburden their innermost thoughts to a pony rather than a person.

Rupert Isaacson's books *The Horse Boy* and its sequel *The Long Ride Home* describe the

The working tack room at the Clwyd Special Riding Centre, North Wales, showing well padded saddles and cage stirrups to hold the rider's foot in position.

extraordinary bond between his severely autistic son and horses. The response to these books gave rise to a new international movement, The Horse Boy Foundation, which specialises in helping children find communication, self-esteem and learning through being with horses.

It seems as though the autistic mind is tuned in to higher octaves of sensitivity. For many persons with autism, communing with the equine mind brings relief from tension and when calm and relaxed their special talents can shine out. In our local RDA branch there was a hyperactive eleven-year-old who babbled incoherently most of the time but as soon as she was on a pony, she could invent stories, often in rhyme, on any subject suggested to her while she was riding.

Although a hug can be such a restorative and reassuring gesture, teachers are no longer permitted to interact physically with children. However, often the first thing a child does on meeting a pony is to give it a big hug and to enjoy its texture and physicality.

Temple Grandin, an American professor of animal behaviour at Colorado State University, was non-verbal until nearly four years old and was ostracised at school for her strange behaviour. Her affinity with horses and cattle has given us insight into the autistic mind as well as insight into the sensibilities of animals. Her lectures, books and films have done much to change social stigma against autism and to promote therapy with horses.

The Greatwood Centre, Marlborough, Wiltshire

Helen and Michael Yeadon began to rescue and re-home former racehorses twenty-five years ago in Devon. The Yeadons shared a keen interest in racing but while everybody was celebrating the winners, Helen was concerned about the then lack of provision for the horses which came in last and were prematurely put down or faced an uncertain future. Since moving to Wiltshire in 2002 they have expanded their horse rescue work to include educational support and therapy for disadvantaged children and young adults.

The connection between the horses and people began when Sophie, the child of a friend, became a selective mute, probably through bullying at school. Sophie and her mother happened to visit the Yeadon's farm just after they had rescued an old mare. The mare was frail and timid and hiding at the back of her loosebox yet when the child stood in her doorway, she came forward and put her nose into the child's hands. There was an immediate affinity between the two which continued to develop and transformed them both. Another early and unanticipated affinity between a horse and a child with complex needs was that of an eight-year-old boy who had twice tried to commit suicide by walking onto the motorway but bonded almost instantly with a particular horse at Greatwood. It gave him a reason for living and restored his self-esteem.

These successes were the inspiration for the Yeadons to set up a scheme in which schoolchildren and young adults with problems could come for weekly visits. This developed into programmes which aim at progression to employment in addition to supporting mental health challenges. Three specialist teachers are now employed by Greatwood to deliver the courses which are fully accredited and complementary to the National Curriculum. Work with the Royal College of Psychiatrists is ongoing and it is hoped that in future, young adults with post-traumatic stress disorder may also be accommodated.

At Greatwood today there are excellent pasture and stabling facilities for about forty ex-racehorses,

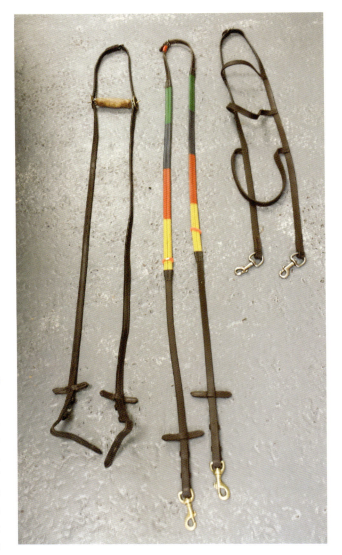

Types of reins: wooden bar (left), rainbow (middle) and ladder (right) adapted to suit the rider. Most of these reins have clips which easily attach to the headcollar or bit. Carriage driving reins may have loops sewn into them with Velcro straps to enable them to be more easily held. Photo by Lynne Munro

most of whom are re-homed. There is no riding here but courses include grooming and stable management for the children who can gain an accredited certificate. For those who find thoroughbreds too daunting, there are Shetland ponies, goats, dogs and hens to look after. A gardening course on which children grow their own produce and learn to cook is also accredited and woodland walks and den building are similarly therapeutic. No child or horse is turned away on the basis of their condition and their success stories are extraordinary. Greatwood is an independent charity and needs £575,000 a year to cover expenses but every penny has a value in providing therapy which is often life changing.

Wormwood Scrubs Pony Centre

At the Wormwood Scrubs Pony Centre (*see* photo page 175) Sister Mary Joy Langdon rescued a young pony called Sioux who was scarred from ill-treatment by a previous owner yet became a similar healing force for disturbed children. Sioux was always totally trustworthy with a child in her loosebox suggesting that when affinity is established it can be mutually beneficial for both horses and humans. It is interesting that horses, especially those who have been mistreated and might be expected to hold a grudge, generally do not seem to be vindictive or to take advantage of the special needs individuals who spend time with them.

The Chariots of Fire Centre, Lockerbie, Scotland

This centre provides specialist carriage driving lessons to improve communication and motor skills in young people and adults. Charlotte, aged twenty-six who has cerebral palsy, began driving with her wheelchair strapped to the carriage and was unable to grip the reins. She progressed to being able to sit on the seat with a waist belt and once the reins were attached to her gloves with a Velcro loop she could take over entirely from her co-driver. Five years later she became the British Novice Para-Driving Champion.

The Fortune Centre for Riding Therapy, Bransgore, New Forest

The Centre was founded by the Hon. Mrs Baillie and Mrs Nelson in 1976; both were RDA volunteers who saw the progress handicapped children were making from just one session a week. They realised that full time courses working with horses could be of even greater benefit, for those aged over sixteen whether they needed assistance or were able-bodied. Their focus was on what these young people could do, rather than on their limitations.

Apart from riding and stable management and the benefits that brings, these students are motivated by the horses to continue their education. In order to care for horses and to gain national accredited qualifications, they can learn literacy and numeracy in an equestrian context and develop IT skills at the same time.

The majority of these students have previously failed to respond to conventional schooling so, instead of a formal classroom, there is a giant interactive screen on the wall connected to the internet. One of the students there was googling the names of different parts of the bridle and relating the names from the image on the screen to a bridle she was holding in her hand, then looking away from the wall and testing her memory. Outside, a young man was using an extra tough all-weather laptop to learn about quantities of feed to give horses which are in or out of work. He had difficulty with keyboards so preferred using a pen to touch the laptop screen to find the information he needed. The incentive to learn more about horses improves the students' reading and number skills to the extent that many progress to being able to lead an independent life and to find work experience and equestrian related jobs.

As Frankie Dettori said when visiting the Fortune Centre, 'Horses are very special, uncritical and non-judgemental. Students can sometimes empathise with a difficult horse and in helping it to improve through patience and kindness, help themselves too.' The Fortune Centre has another claim to fame; this is where Carl Hester (*see* page 177) was first employed, aged sixteen, as a working student for riding and stable management and where he began his international dressage career.

The RDA and centres offering equine therapy have been a spectacular success. Their flexible approach in improvising and adapting the tack and equipment to a range of special needs is achieving real clinical progress as well as providing all-important motivation for everyone involved.

CHAPTER TWENTY-NINE

The Royal Mews

In the heart of London, the Royal Mews borders the gardens of Buckingham Palace. Here the Crown Equerry is in charge of the horses, carriages and cars which transport the Royal Family and their official visitors. Thirty horses and about eighty carriages and coaches are housed at the Royal Mews together with one of the finest collections of working harness in the world. Entrance to the Mews is through an imposing stone arch into a large quadrangle completed in 1825 by John Nash and surrounded by elegant two storey buildings with staff accommodation above the stabling. A group of mature London plane trees in the centre and glass canopies outside the carriage houses make this an attractive and practical working space which is constantly busy.

The word 'mews' has its origin in falconry. During the annual moulting phase of a falcon it is rested from work until its new plumage is complete. During that time it is caged and makes a characteristic mewing sound. The caged falcons were sent to the stables where they could be looked after alongside the horses until they had returned to their working condition.

Whereas equestrian disciplines have their own individual histories, the Royal Mews carries the history of Great Britain. The state occasions of royal marriages, funerals and anniversaries and official visits from foreign dignitaries are part of our constitutional and political affairs and are distinguished by ceremonial carriage processions.

The Royal Mews horses are involved in the State Opening of Parliament, the Queen's Birthday Parade, the Garter Ceremony, Royal Ascot and visits by heads of state. Newly appointed high commissioners and ambassadors are transported by carriage to Buckingham Palace to present their credentials to HM The Queen. At least twenty-five countries are represented in this way every year and reflect the international duties of the Royal Mews.

Gifts to the Sovereign such as the state coach presented by Australia for the bicentenary in 1988 through public subscription, and the black mare Burmese from Canada, ridden side saddle by the Queen for the Trooping of the Colour ceremony until 1986, illuminate countries across the world which have links with Britain.

In 2017 there was an enquiry from Hawaii, regarding the restoration of a carriage thought to have been commissioned in England by Queen Liliuokalani and then exported. It relates to an audience she had with Queen Victoria in 1881 when staying with the Prince of Wales and reflected the English influence abroad.

Those who work in the Royal Mews form a close community and many of them live on site. Generations of some families have worked for the Palace – Martin and David Oates are the fourth generation of their family to be employed here. The carriage restoration and the saddlery and harness work are carried out by master craftsmen at the Royal Mews. Other maintenance is contracted out to wheelwrights, heraldic artists and tailors for livery, carriage upholstery and embroidery work.

The last carriage to be commissioned by the Sovereign was in 1902, made by Hooper & Co. for Edward VII and in 1910 the first automobiles were

Inspection of the carriages about to convey ambassadors to Buckingham Palace to present their credentials to Her Majesty The Queen.

garaged at the Mews. Queen Victoria is quoted as saying to her Master of the Horse, the 6th Duke of Portland, 'I hope you will never allow any of those horrible machines to be used in my stables.' Ironically, it was a company (Forder) holding the Royal Warrant for supplying carriages to the Prince of Wales in 1875 which provided the coachbuilding expertise for the early Rolls-Royce cars. Now cars and horses are stabled together harmoniously, though even a Rolls-Royce Phantom has to give way to the horses within and beyond the Royal Mews.

The attention to detail and authenticity of the equipment is evident everywhere. For example, the underside of the folding steps of the carriages will only ever be seen by the footman who operates them, yet each is exquisitely decorated in gold leaf, the horses' day rugs each bear a hand-embroidered royal cypher, and the harness on the life-size model horses drawing the Gold State Coach is all genuine.

The original blue George III state harness was first used in 1762 for the State Opening of Parliament. The early 19th century Red Morocco state harness shown overleaf (with breast collars instead of the usual neck collars), was worn by eight horses to draw the Gold State Coach which weighs over four tons. It was used at every coronation since and finally for the last time, for the Golden Jubilee procession of HM The Queen in 2002, a remarkable span of about two hundred years.

The harness is now considered too fragile to be used safely, though still looks magnificent. The quality of the leather and fine workmanship is shown by the stitch count of fourteen or fifteen to the inch; modern leather can usually not support more than twelve stitches per inch otherwise it may perforate. Conservators look after the state harness to prevent any deterioration.

The proximity of leather and metal in saddlery and harness has always caused difficulty when being cleaned. Liquid metal polishes can damage the leather and stitching and many also diminish the metal buckles or terrets to which they are applied by removing a fine layer of metal each time.

Separating leather from metalwork as far as possible for cleaning is a good reason for first dismantling saddlery and harness completely, though buckles cannot always be undone from the leather turn they are sewn to.

The most harmful ingredient in many polishes is ammonium hydroxide and when used on plated metal soon reveals its base and erodes patterns and engraving. Irreversible damage may be done to brass, a porous material, which can result in a pinkish bloom instead of a bright true shine. The method currently recommended by museums and conservators is the use of dry cloths impregnated with polish which are non-abrasive, do not rot stitching and bring up a long-lasting shine.

Because of the quality and rarity of the carriages and their accoutrements there is nothing ordinary at the Royal Mews and so an extraordinary

The Red Morocco state harness is thought to date from the turn of the 18th/19th century. It is adorned with most of the original decorations used on the blue state harness commissioned by George III in 1762. 'Red Morocco' refers to the leather which is goatskin tanned with sumac and the buckles are of twenty carat gilt ormolu. The postilion's saddle is unusual as it has no skirt, the stirrup bar is under the flap. The collar (top) was used to draw the gun carriage for Queen Victoria's funeral and the gilt boss is reversed as a sign of respect.

Detail of the saddle pad of George III's Red Morocco state harness. The coat of arms of the United Kingdom with the lion and unicorn dates back to 1603. George and the dragon are also depicted and every buckle has a crown. The tooled pattern of olive leaves on the pad strap matches the border on the postilion's saddle and is perhaps a symbol of peace following the Seven Years' War. The straps are edged in black leather piping and finely hand stitched throughout.

The State Harness Room at the Royal Mews, designed by John Nash in 1825. A bridle with ornate brasswork can be seen on a stand (left) and a High Commissioner's carriage, drawn by Cleveland Bays with postilion, is passing the doorway. Harness is protected inside glass cabinets the length of the room.

team of experts is needed for their maintenance. This is not only for the preservation of these historical pieces but to ensure that the knowledge of craftsmen such as saddlers and coach restorers is not lost. The Royal Mews takes the responsibility of passing down these unique skills to the next generation through apprenticeships.

Master Saddler Frances Roche has been at the Royal Mews for twenty-one years and is the first professional harness-maker to work on site. She is assisted by Master Saddler Catrien Coppens, a former student of hers from Cordwainer's College (now the Princess Royal College of Animal Management and Saddlery in Enfield, London), and an apprentice. The apprenticeship here, which includes carriage restoration and saddlery skills, is sponsored by the Worshipful Company of Coachmakers and Coach Harness Makers and indentured on the Worshipful Company of Saddlers' Millennium Apprenticeship.

Together they repair and remake all the harness and saddlery at the Royal Mews and Windsor Castle Mews to keep it working safely and its appearance excellent.

The workshop is a spacious modern room lined with benches. It immediately stands out from other harness workshops by the quality and decoration of the harness here for repair and by the lack of sewing machines.

A large central table is spread with parts of bridles and harness in the process of repair. Hand tools such as a pricking iron to mark the number of stitches per inch, an edge shave to reduce the thickness of the leather, an awl and linen thread, one of the strongest natural fibres known, are a world away from automated machinery. Two 'stitching horses', pieces of furniture with a seat and integral saddler's vice known as a clam, are in use by the two saddlers.

The amount of harness at the Mews is considerable; the principal George III and William IV Morocco sets were each made for eight horses. The harness and bridle (a 'side') for each individual horse weighs about fifty kilograms (110 lb). The bridle alone weighs eight kilograms (18 lb) and is attached to a 'neck lay', a network of decorated straps which lies over the top half of the horse's neck and is thought to help counteract the weight of giltwork on the rest of the bridle.

In addition there are seven sets of black state harness. Each of these sets is made up of sides for six horses and dates from the 19th century. Young horses are trained in lighter harness and gradually learn to wear the state harness. Although heavy, it is not worn for very long on each occasion. There is also harness for schooling and exercising and for the Brougham which conveys the daily internal Palace mail.

Harness in constant use is subject to wear and tear, perhaps particularly so here because of its unusual weight. Also the amount of decorative metalwork can damage leather through abrasion if the harness on one horse should rub against its neighbour's harness.

Bridles with Buxton bits are fitted to the individual horses, as are the collars and each horse also has its own fitted saddle. The harness is adjustable on both sides for extra comfort. Collars and hames are made by John McDonald, Master Collar and Harness Maker, in Somerset but repaired here.

All state harness is stitched by hand, showing a finer quality of work at twelve stitches per inch (and traces at ten stitches per inch), which is the strongest and most enduring method. Similarly the semi-state harness or Ascot Harness is stitched by hand at ten stitches per inch (except for the traces and breeching which are stitched by machine at eight per inch). Traditional materials, such as waxed linen thread, are used and repairs are sensitively carried out so that as much of the original piece of harness as possible is preserved; conservation and restoration are finely balanced.

Buckles, the shape of which, Fluted, Round, Square and Council, define the particular harness, are repaired and re-gilded by jewellers outside the Royal Mews.

Her Majesty The Queen travelling in the Irish State coach for the Opening of Parliament.

The Royal Family and Horses

The stable staff start at 6am and after feeding and grooming the horses they exercise them in the indoor Riding School or outside in the roads and parks of the city. The generous proportions of the Riding School, built in 1764 by William Chambers for George III, provide an essential quiet space for schooling and exercising. Queen Victoria's many children all learnt to ride here.

Many of the horses, Cleveland Bays and Windsor Greys, are bred at Hampton Court and these are named personally by Her Majesty The Queen.

Each is treated as an individual and this is exemplified by a plaque in King Edward VII's Town Coach used to convey Lady Churchill and her two daughters for the state funeral of Winston Churchill on 30th January 1965. The names of the two bay horses which drew the coach that day, 'Capetown' and 'Bloemfontein', are engraved on the memorial plaque together with that of the Coachman driving them.

One of the Queen's favourite Windsor Greys, named Daniel, was retired in 2017 to the Horse Trust in Buckinghamshire after fourteen years of service. Daniel also featured in the Royal Mail stamp collection celebrating working horses and is commemorated by a life-size statue, commissioned by local residents for the Diamond Jubilee, in Windsor.

Horses at the Mews return every year to Hampton Court to be turned out to grass on holiday from their official work.

Her Majesty The Queen has adopted the Cleveland Bay, our oldest British breed, for carriage work and through her breeding programme has rescued it from decline.

In the 1960s it was on the verge of extinction when the Queen purchased a colt (Mulgrave Supreme) about to be exported. She made him available at stud and the versatility of this breed has since established its success for both riding and carriage driving. This is just one example of the interest and participation of the Royal Family who stand out in the British equestrian landscape as a beacon of sportsmanship and patronage across an impressive range of equine disciplines.

The Queen has a deep knowledge of the breeding and training of flat and National Hunt racehorses, some of which are bred at the Royal Stud at Sandringham. She is Colonel-in-Chief of all the Guards regiments and has been a Patron of the British Show Jumping Association since 1952. Besides breeding Cleveland Bays which are categorised as an Endangered breed, the Queen has a Highland Pony stud at Balmoral where ponies are bred for riding.

The Princess Royal has represented Britain in the Olympic eventing team in Montreal and won a gold and two silver European Championship medals. She has also hunted, steeplechased and won on the flat at Ascot. Her many official equestrian positions include being an ex-president of the International Equestrian Federation (FEI), President of World Horse Welfare, Patron of Riding for the Disabled and the Injured Jockeys Fund, Honorary Master of the Worshipful Company of Saddlers and Past Master of the Loriners' Company and Colonel of the Blues and Royals Regiment. Her daughter, Zara Tindall, was the European Eventing Champion and won an Olympic team silver in London in 2012.

Prince Charles is an experienced rider. He has also championed the working draught horse such as the rare breeds of Suffolk Punch, Highland and Dales ponies and is Patron of the British Horse Loggers.

The Duke of Edinburgh is a founder of British competitive carriage driving and regularly competed internationally, as well as having been a keen polo player. He introduced the Prince Philip Cup Games in 1957 which gave children an opportunity to compete in teams, regardless of what sort of pony they had. The Prince Philip Cup national finals are held at the Horse of the Year Show and continue to be very popular with both riders and spectators.

The Royal Family carry out hundreds of public engagements every year and are conveyed on these

Frances Roche, Master Saddler at the Royal Mews for twenty-one years. She is hand-stitching a noseband held steady in a saddler's clam. On the bench in the foreground is a newly gilded decorative boss for the bridle of the Master of the Horse.

The bit belonging to the Master of the Horse's bridle. Detail of a handstitched decorated throat lash in dark blue leather.

occasions by the Rolls-Royce and Daimler cars at the Mews. However, a fleet of cars could not evoke the historical events that have shaped this country in the same way that the ceremonial splendour of horses and carriages does so well.

The horses, which belong to the nation rather than the Queen, remind us of our dependence on them in the past for transport, war, commerce and agriculture. The splendid liveries of the outriders, coachmen and postilions faithfully represent those worn in earlier centuries and the symbolism in the decoration of the coaches all show how much our nation's history is valued.

The Diamond Jubilee State Coach, built by Jim Frecklington, is an outstanding example of historical representation, and could perhaps be thought of as a working time capsule of great beauty.

It was first used for the State Opening of Parliament in 2014. Incorporated into its interior coachwork are samples representing past campaigns, exploration, science, architecture and tradition, such as a fragment of iron from the Sebastopol cannon from which Victoria Cross medals are cast, a flake from the Stone of Scone which has been present during the crowning of our monarchs since 1308 and a shard from a lead musket ball from the battle of Waterloo. Samples of wood from the Arctic bases of Captain Scott and Ernest Shackleton, Henry VIII's flagship Mary Rose, Caernarfon Castle and Durham and Canterbury cathedrals are just some of those which line the inside of the coach doors. In contrast to the history, the structural technology of the coach is contemporary with carbon fibre brakes, a hydraulic anti-sway mechanism, electric windows and air conditioning.

HM The Queen personifies duty, continuity and stability. A royal procession, immaculately turned out by the Royal Mews, celebrates these ideals with colour and vivacity and is internationally admired. It is also an expression on behalf of the nation of our individuality in the world.

CHAPTER THIRTY

Show Jumping

Few would have realised the significance of a summer's day at the Richmond Junior Show Jumping Championship in 1939 when three children shared the first prize. They were Douglas Bunn, Pat Smythe (both aged eleven) and Fred Winter (aged thirteen), each of whom would leave their mark in equestrian jumping history at home and abroad.

Fred Winter (1926-2004) went on to break records as a jump racing jockey and trainer. He was the only man to have won the Grand National, the Champion Hurdle and the Cheltenham Gold Cup both as a jockey and as a trainer. Besides this, he is perhaps best remembered for his ride on Mandarin in the Grand Steeplechase de Paris in 1962, a race as daunting as the Grand National. Despite being weak from a severe stomach upset, the bit coming apart at the fourth fence (of thirty) leaving Fred with no control, and a serious injury to Mandarin's foreleg three jumps from home, he just held on to his lead to win by a head. It is recorded as one of the ten best races of all time.

Pat Smythe (1928-1996) became a household name riding Tosca and Prince Hal at a time when show jumping commanded a television audience of millions. She emphatically demonstrated that women could successfully compete and win against men (as well as receiving several marriage proposals from them) at the highest international level. Women were banned from taking part in the Nations Cup team until 1952 and in the Olympic team until 1956. Pat was the first woman to both be in a British show jumping team and win an Olympic team medal, a bronze in Stockholm in 1956. Later she was the team's chef d'équipe and President and Vice President of the British Show Jumping Association (BSJA).

Douglas Bunn (1928-2009) was a pioneer too and a flamboyant character, never wasting a minute of a life packed with business and family, socialising, competing and innovating. Winning a scholarship to Cambridge, he qualified as a barrister but soon returned to his real interest which was riding and show jumping in particular.

Hickstead and the birth of modern Show Jumping

In the early days after the war, British show jumping was male dominated and influenced behind the scenes by those who had been in the armed services. After distinguished war service, Colonel Harry Llewellyn stood out from his colleagues and decided to resume his equestrian interest (having come 2nd in the Grand National in 1936) by trying his hand at show jumping. His search for a horse was narrowed down from twelve horses registered with the BSJA to two. One of the two, Foxhunter, whinnied as he approached them and was duly chosen. Llewellyn and Foxhunter went on to become national heroes jumping for Britain thirty-five times, the highlight being to secure Britain's only gold medal at the Helsinki Olympics in 1952. That same year HRH The Queen became patron of the BSJA. After an accident in his field, Foxhunter died aged nineteen and was buried under a handsome memorial near Abergavenny. Sir Harry's ashes were later scattered on the grave.

Regardless of the brilliance of individuals such as Harry Llewellyn and Pat Smythe, show jumping was generally a lacklustre sport in Britain until the late 1950s when Douglas Bunn lit a fuse under its hidebound conventions. He independently established Hickstead as the centre of British show jumping by building a course that was the first of its kind in Britain. Despite much opposition, his vision and determination was responsible for its success today and for our show jumpers becoming competitive abroad, for which he earned the title 'The Master of Hickstead'.

Douglas was a member of the British team in 1957 and saw how the British and European approaches to show jumping were diverging. The British courses were designed to be jumped carefully and accurately over poles which were easily dislodged. In contrast, the European courses were big and solid but inviting and demanded a bold, more exciting and freer style of riding. In 1959 Douglas had bought a Tudor farmhouse with a hundred acres at Hickstead, West Sussex, on the old coaching road between London and Worthing. His vision was to establish a permanent showground with an unrivalled jumping course in continental style. This he achieved with the All England Jumping Course which opened in 1961.

The British Jumping Derby has been an internationally renowned event for at least fifty years over this course which includes the notorious Derby Bank. Douglas had seen a three metre high bank in Hamburg, an obstacle on a course so difficult

Dereck McCoppin on Whiterock Lucky Lady flying the water jump in the All England jumping Championship at Hickstead in 2015.

Shane Breen on Acoustic Solo du Baloubet jumping expertly off the Hickstead Derby bank in 2015.

that the first clear round for fifteen years was by the course designer, Herr Pulverman. Marion Coakes (née Mould) on Stroller was another winner there and regarded this as an achievement equal to her Olympic medal win. Douglas obtained permission to go to Hamburg to examine and measure the jumps. Arriving in a snowstorm on New Year's Eve, there was nobody to meet him, everyone having assumed his visit would be abandoned due to the weather. Douglas hurried to the show ground before the jumps were blanketed by snow and returned with all the measurements to build his own, even bigger version at Hickstead. Together with the course designer, Pamela Carruthers, they added a unique and varied course of open ditches, oxers, water jumps, post-and-rails, stone walls and hedges. Today the Derby Bank stands at 10' 6" high with the main sloping face at 60° to the ground.

Horses are required to jump a fence on top of the bank, descend the slope and in two strides jump a high upright fence.

Douglas was ahead of his time and interest from both competitors and sponsors was dismal to begin with but gradually British show jumpers adapted to the change, encouraged by good prize money and enthusiastic competitors from Europe and America. Since then the idea of a Jumping Derby has been copied round the world.

The year Hickstead opened in 1961, the FEI chose it as the venue for the Junior European Championships and it soon became an internationally important centre. Not content with all this, Douglas was the first to establish Team Chasing where teams of four compete round a cross-country course. The popular working hunter class was another of his innovations, an arena-based event which includes natural obstacles and opening and shutting a gate.

Douglas would also fly his helicopter to Leicestershire to have a day's hunting with the Quorn as well as being Master of the Mid-Surrey Draghounds, but he kept his finger on the international pulse and was influential in the appointment of Ronnie Massarella, one of the most successful chef d'équipes to the British team, a post he held from 1969 to 2000.

The original Hickstead Place stable yard of traditional Sussex brick and tile has seen many famous horses such as Beethoven, Douglas Bunn's British Team show jumper. Olympic medallist David Broome also rode him to become World Champion in 1970. Beethoven endeared himself to spectators with his characteristic heel and tail-flick over every jump and attracted the attention of Pope Pius XII. When jumping in Rome, Douglas was called to the Vatican for a private audience where he reported that the conversation between them was more about horses than religion.

Pat Smythe's grey horse, Tosca, was also a guest at the yard. When her show jumping days were over she lived in peaceful retirement on a farm in the Vale of Evesham where, incidentally, the radio series of *The Archers* first began. Another visitor was the incomparable Stroller, owned and ridden by Marion Coakes. Stroller arrived from Ireland as part of a job lot and, although barely 14.2hh, went on to represent Britain for nearly twelve seasons with numerous successes, including an individual silver medal in 1968 at the Mexico City Olympics.

Breen Equestrian Ltd at Hickstead Place

Douglas Bunn's daughter Chloe, her husband Shane Breen and their children now live in the family home where they have established Breen Equestrian Ltd. From here they train and send out top level show jumpers and are one of the top three breeders of show jumpers in the country. Shane, from Cashel in Tipperary, has been competing internationally for over twenty years and has ridden for Ireland since he was a junior. Inspired as an eight-year-old spectator with his father at the Dublin Horse Show, he watched the tense jump-off for the Aga Khan Trophy as Eddie Macken clinched the Nations Cup victory for Ireland – and chose to be a show jumper rather than a jockey.

Shane's schedule is non-stop travelling, training and competing, apart from a rare weekend at home and a little hunting in Ireland. When I met Chloe and Shane, he had just arrived back from a successful visit to Gijon, Spain but was due to jump in Switzerland a few days later before returning to Barcelona. The horses travel well by road to Europe. When competing in Hong Kong or Dubai, they will be rested for forty-eight hours before going into the ring though don't appear to suffer from jet lag.

Beyond the original Hickstead Place stable yard, there is further stabling to accommodate about forty horses in all, a covered horse-walker with central lunging ring, an indoor school and an American barn with a very fine tack room.

Historically, the rider's forward seat and shorter stirrup leathers for jumping, introduced by Fedrico Caprilli, led to a departure from previous saddle designs which had been more suitable for dressage. The saddle flap was extended forwards to accommodate the rider's knee and continental saddlers, such as Toptani and Hermès, designed the classic close-contact jumping saddles which have become a standard pattern.

Saddlery made for jumping is well-established in Europe and Shane finds Bruno Delgrange saddles the best balanced and most comfortable for himself and his horses. Made in France but with English leather, the saddles carry the distinctive Master Saddlers' round knife motif with the BD initials on

The late Douglas Bunn's tack room at Hickstead with its solid fuel stove which is still working today.

either side. The saddles are of a conventional jumping design with a regular tree, pommel and cantle, and a forward cut flap underneath which are small blocks in front of and behind the rider's leg. Most of Shane's saddles are made to measure for his horses and Bruno is an expert fitter being a rider as well as a Master Saddler. A protective leather pad (stud guard) on the underside of the girth prevents the studs in the shoes from grazing the horse when its forelegs are tightly flexed over a jump.

With the introduction of speed jumping, saddle design became more supportive to the leg to accommodate sharp turns and change of pace and riders can have customised blocks built in. Stirrup design changed too and Shane's preference is for Flexon stirrups which have a broad angled-up base with plenty of grip. His bridles are also made by Bruno Delgrange with bits to suit the individual horse. Occasionally, some horses jump better in hackamores, such as Eddie Macken's outstanding Boomerang who won the Hickstead Derby for three consecutive years.

Breeding and training show jumpers

When I ask about the different breeds, Shane explains that the show jumping stud books show pedigrees of mixed origins. Due to the availability of frozen semen, continental and Irish crosses are common. Long ago, a touch of Clydesdale would have been permissible (as in Foxhunter's breeding) but now Thoroughbred genes help to meet the high standards and requirement for speed. 'Ideally', says Shane, 'if a Thoroughbred sprinter stallion such as Yeats, with his powerful physique for speed, were to cover a proven show-jumping mare, they could, in theory, produce the perfect show jumper. Unfortunately, the top racing stallions are restricted to covering thoroughbred mares which are DNA typed and registered with Weatherbys in the Stud Book.'

Being good at recognising a promising horse is a special talent, one that Shane picked up from Tommy Wade, the show jumper and dealer, with whom he worked as a teenager. Tommy had an eye for a horse having spotted the 15.2hh Dundrum who used to pull the luggage cart from Dublin Station to a hotel. They became the toast of Ireland when Dundrum cleared 7' 2" at Wembley in 1961.

Shane described how he brings on youngsters: 'It's possible to tell very early on if a horse has potential. I free jump them as a yearling and look again when they are two and three years old. Some love to jump, others don't, some tend to jump with a flat back, others make a nice arch (bascule) over the jump.

'Starting with a horse that has the natural balance, boldness and athleticism to become a Grand Prix winner makes a big difference. By contrast, you

A superb tack room at Breen Equestrian. The room is panelled throughout with a full length polished counter of 3" thick, home-grown Hickstead oak. Behind Shane are his Bruno Delgrange saddles and above are some of his many trophies. Drawers and cupboards contain folded rugs and numnahs.

The stallion barn at Breen Equestrian with Golden Hawk being led in for grooming. On the modern Musgrave-style mobile saddle horse is a Bruno Delgrange saddle. Sheepskin lined girths are hanging up behind in the solarium bay.

can only improve a mediocre horse so far and it will never be a great horse. By "improvement" I mean that with careful training and, most important, a rider who can give the horse the confidence it needs, its performance can be enhanced before it reaches its limit. It's where that limit is set that is key.

'Recently, I turned out four weanlings into a new field that hadn't been topped and there was a band of thistles between the gate and the rest of the pasture. They all dashed off, three of them jumping over the thistles like stags. The fourth ran straight through them and he had been bred from the least successful show jumping mare. In my opinion, probably 75% of the foal's talents come from the mare.

'Our promising young stock are broken in when rising four years old and then put away for a year. When they're six to seven years old they should be competing amongst their own age group and jumping one metre twenty. After that, through lots of flatwork, they continue to develop their suppleness and agility and learn to shorten and lengthen their strides – which is crucial for positioning when approaching a jump. And, just like elite athletes, keeping up their fitness levels gives them a long working life.'

Hickstead, with its superb facilities continues to be a vital proving ground for British team members and is a venue for many disciplines, such as driving and dressage. Inclusive competitions for all levels of rider, including Pony Club members, take place here as well as top level events such as the King George V Gold Cup and the FEI Nations Cup of Great Britain at The Royal International Horse Show.

Without the Master of Hickstead's foresight and determination, the roll call of some of Britain's show jumpers such as David Broome, Liz Edgar,

Scott Brash, in the Hermès Grand Prix outlined against the glass ceiling at the Grand Palais de Paris. He won an Olympic team gold medal in 2012 and is the only rider to have won the Rolex Grand Slam by winning the Masters Grand Prix at Spruce Meadow, Geneva and Aachen in 2015.

Harvey Smith, the Whitakers, Nick Skelton and their many successors such as Ben Maher, Laura Renwick and Scott Brash, might have looked very different. Douglas Bunn is commemorated by Philip Blacker in a striking relief situated next to the International Arena depicting him jumping a wall on Beethoven.

How delighted he would have been with the team gold medal for show jumping in the London 2012 Olympics and Nick Skelton's Puissance world record (2.32m or 7' 7.3") on Lastic and his 2016 individual Olympic gold in Rio on Big Star.

Show Jumping Terms

Oxers: a show jump with parallel bars.

Hackamores: a bitless bridle which can exert pressure on the nose and poll of a horse instead of the mouth.

Blocks (on the saddle): solid leather-covered blocks of various shapes, sizes and positions built into a saddle to support the rider's leg.

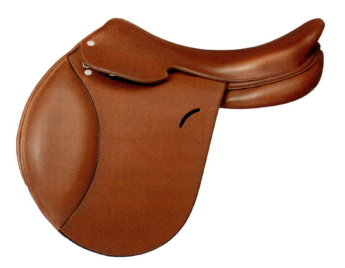

Hermès Cavale saddle, a classic, close-contact show jumping model with padded flaps, wide tree, and integrated panels.

CHAPTER THIRTY-ONE

Deer Stalking

Atholl Estate, Blair Atholl, Perthshire, Scotland

In the dramatic scenery of the Grampian Mountains, tradition is alive and well. Highland ponies are still used to carry the deer off the hill on the Atholl Estate where the Red deer herd is scrupulously managed. Tweed, willow and leather remain the basic materials for traditional stalking and are used in the clothing, panniers and saddles, exactly as they were when Mary Queen of Scots stayed here at Blair Castle, Perthshire, in 1564.

Just below the tree line where the peaty waters of the River Tilt tumble over boulders is a shooting lodge with outbuildings. Here the tack room, one of several on the estate, is a simple stone building. It has a long history and over the decades has been used for a variety of purposes. Its rough walls are patchworked with brick where the original bothy has been repaired and the roof raised to make a single bare-raftered room for the stalking and grouse shooting tack.

A flight of worn steps leads up to a large wooden door in the lee of the prevailing wind. Inside, as your eyes adjust to the gloom, you can see several pairs of willow panniers stacked in a corner and wooden saddle racks set into the stone walls. A ladder propped against the wall makes a convenient set of shelves. Harsh winters and demanding work are reflected in this essentially practical store for the equipment and little time is spent in it. Most of the tack cleaning takes place in the warmth of the back kitchens of the lodge nearby.

The panniers are works of art in willow and require such skill in the shaping of the basket using a mould, that the experts who can still make them are now few and far between. The paired panniers are placed over the deer saddle and during the grouse shooting they are used to carry the lunch out to the hill and the shot birds back. Light and capacious with a canvas cover on the lid and plenty of ventilation through the decorative weave, they are excellent for portage, provided each is carefully balanced.

The heavy-duty deer saddles, at least sixty years old, are regularly oiled and in good condition. They are longer and flatter than riding saddles and have no stirrup leathers. The tree is of wood and ridges at the front and back help to keep the deer carcase in place. Instead of being stuffed with sheep's wool, the saddle padding is made from a coarse straw or reed-like material mixed with horse hair and covered in wool collarcheck so that it dries out quickly. This is important in a region that receives over 1,500mm of rain every year. Gervase Markham, a saddler in 1607, favoured deer hair for stuffing, though, 'for it is softest, leyeth moste even and soonest dried when it is wette'.

A more modern combination saddle remains stored in the tack room. Its design for both riding and carrying deer was not a success, being too uncomfortable to sit on for long distances and unsatisfactory for carrying deer.

The deer saddle is secured with three girths, and sometimes a crupper, to prevent the deer carcase from slipping and causing a hazard to the pony on precipitous slopes. A wide webbing girth is used as a breast strap to prevent the load from sliding backwards and another, held up by hip straps and attached to the back of the saddle flaps, acts as breeching around the hindquarters to stop the saddle from moving forwards.

Stalking requires great skill to achieve a cull without distressing the deer herd, as well as careful training to manage the ponies for the retrieval of the carcases. The hillmen guiding the stalkers will be thoroughly familiar with the terrain and hefting habits of the local herds.

It is essential to stay downwind and out of sight of the deer as they have acute powers of smell and sight. The intention is to bring down an animal with a single shot in an area where the pony can retrieve the carcase without danger from bogs, slippery scree or steep slopes.

Once the animal has been shot and killed outright, the stalking party remains hidden until the herd has moved on and continues to graze. This is to ensure minimum alarm to the herd. The deer is then gutted (gralloched) and drained of blood. This must be drained promptly; if the blood is allowed to congeal it taints the meat with an unpalatable musty flavour.

The ponies, waiting some four hundred yards behind, are then brought up and the deer carcase is corded onto the saddle by three leather straps. A surcingle (overgirth) runs right round under the pony and over the saddle. It is buckled on top with the long end left in a loop beyond the buckle tongue for easy unfastening as a safety measure in case the pony should sink into a bog and have to be unloaded

Stalking tack room at Forest Lodge, Atholl Estate. The stone walls of this utilitarian tack room show their age through many repairs and the panniers and saddles reflect a way of life unchanged in centuries.

Highland ponies, Juniper and Diridh, take a break bringing a deer off the hill.

quickly. Care is always taken when loading a stag onto the saddle to secure its head and antlers as far as possible out of the peripheral vision of the pony's sight and bridles with blinkers can be used if necessary. A mature stag may weigh about fifteen stone (95 kg) but the Highland pony is sturdy and surefooted.

The ponies usually wear headcollars rather than bridles and when working as a pair, the long rope of one is passed through a saddle terret of the pony in front so that the keeper may lead both in single file on steep tracks. Heels on all four shoes prevent the pony from slipping and sliding on snow or mud. However, if the stalking party becomes lost in the mist, giving the pony its head, holding onto its tail and trusting its homing instinct is the quickest way to get back to a hot bath and a glass of whisky. Atholl Brose is a famous whisky punch which was used in 1475, not just as a restorative but as a weapon of war. The then Duke of Atholl filled the well of his arch enemy, the Earl of Ross, with Atholl Brose and after drinking deeply the Earl was easily captured. Paintings of deer hunting by Victorian artists such as Sir Edwin Landseer (*Monarch of the Glen* was painted in the Grampians in 1851) show that deerhounds were often used until the middle of the 19th century to pursue wounded stags.

Today's stalking is quite different; no dogs are used and legislation ensures humane conservation. The increase in deer populations led to the Scottish Deer Act in 1959 and the English Deer Act in 1963. These Acts protect the deer during their breeding season and specify the calibre of rifle to be used for culling. Telescopic sights are used for accuracy and stalking clients have to prove their competence by target shooting on a range before being allowed to participate.

The deer have to be culled selectively every year to maintain the health and quality of the herd and to avoid exhausting the food resources. Since wolves became extinct in the late 17th century, deer have had no natural predators and numbers can quickly outstrip food resources. Old animals, and calves born after July, are likely to die anyway from exposure during the winter. Gestation of Red deer is approximately 232 days and, in these harsh conditions, it is common for calves to be born only every other year.

Management of the deer population is a complicated issue with significant ecological and economic impact. In total, some 2,500 full-time jobs in Scotland depend on deer management which is estimated to be worth £148 million per annum (from a survey commissioned by the Association of Deer Management Groups in 2016). Strategies for balancing the often conflicting needs of deer welfare, forestry and agriculture, tourism, and venison production are determined by a number of organisations such as the Scottish Parliament, Scottish Natural Heritage, the British Deer Society and the Scottish Gamekeepers' Association. Red deer populations are estimated to be mostly stable while those of roe deer are rising, in spite of losses during the severe winter of 2009–2010.

The total number of deer in Scotland (red, roe, fallow and sika) is currently very approximately 750,000.

Blair Castle on the River Tilt, the home of the Dukes of Atholl for seven centuries, was built to defend the mountain passes to the north. Queen Victoria gave the family the right to command a private army, the Atholl Highlanders, which is still retained to the present day and is the only private army remaining in Europe. Full Highland dress in Murray tartan is worn on parade and the 12th Duke of Atholl wears a golden eagle feather in his glengarry cap to signify that he is the chieftain.

Tweed suits (a long jacket and plus-fours) and caps are worn by the estate hillmen such as Graeme Cumming, the head stalker, and beat keeper Richard Fraser. The tweed is woven in the Atholl Estate colours in order to blend with the palette of the Grampian mountains; lead blue for the River Tilt, white for the marble and quartz veining in the granite and the limestone, and brown for the peaty soils. Spun by Haggarts of Aberfeldy from Scottish wool, the tweed is practical, beautiful and good camouflage. Scottish Blackface or Cheviot fleece with its high oil content is chosen for natural weatherproofing and

Stalking saddle detail. Spare snaffle bits are hung on the saddle ready to be attached to the headcollar if more control is needed. Blair Hamilton, in apprentice's tweeds, is quite comfortable without gloves in a temperature of -4°C on this day.

On the hill. Two culled stags loaded for home on a fair day in the Grampians. Atholl Estate, Perthshire.

tightly woven to be windproof. More than a hundred Scottish estates hold copyright of their tweeds with Haggarts who have been in business since 1801.

The estate of some 145,000 acres is traditionally managed with sheep and cattle at lower altitudes. So keen was the 4th Duke of Atholl in 1790 on the productivity of his land that he had a special medal struck by the Royal Mint, engraved 'God Speed the Field' as a prize for a ploughing match. Today, conservation is an important goal and has encouraged native species to thrive such as the red squirrel, eagle, capercaillie and black grouse.

Glen Tilt resulted from an unusual geological fault, giving limestone to the east of the river and a mass of granite to the west. The ecology is therefore unique with many rare lichens and mosses. Limestone is extremely important to grazing animals and the difference in diet between the east and west slopes is reflected in the characteristics of the deer herds.

Of the 8,000 Red deer on the estate, those grazing on the eastern side of the river have a significantly higher bodyweight and larger antlers than those living on the western side. Antlers with six points on each side (Royal) or with seven each side (Imperial) are not necessarily more common in older animals but signify dominant individuals.

On the Atholl estate, the thirty Highland ponies have pedigrees going back to Victorian times. A short-legged, broad-backed pony of 13.2 to 14.0 hands high with a good temperament is the ideal type. Tough and attractive, these ponies excel on the mountain terrain, though caterpillar-tracked, amphibious vehicles, such as the Argocat, are more suitable for very marshy areas. In the summer, pony trekking is popular with tourists. This varied work keeps the ponies fit all the year round and contributes to the economy of the estate. Foals accompany the mares while working on the hill and so become accustomed to the sights and sounds of stalking. Deer skins are also hung over their paddock fences so that the young ponies in training become used to their smell.

To enter the realm of the deer, the stalker must measure himself against their vigilance, the wildness of the mountains and the mercurial weather. He must be tough and wily to succeed. Such immersion in the

Blair Castle, Perthshire. Highland ponies are used for deer stalking in the winter and trekking in the summer. The World Piping Championships and Blair International Horse Trials are also held here.

natural beauty of the Highlands can be an intensely spiritual as well as a physical experience. Traditional deer stalking remains the most sympathetic method of balancing the herd numbers with the grazing resource and of maintaining the genetic quality of the deer herd.

The sustainable management on the Atholl Estate is an example of harmony between man and nature. Thus the deer survive to their best advantage in a natural mountain habitat, the Highland ponies, bred for working, thrive and the skills of the hillmen continue to be passed on for the future.

Deer Stalking Terms

Tree – is the internal frame of a saddle made of metal, wood or synthetic material.

Collarcheck – is a woven, 100% wool material with a traditional check pattern used particularly for the underside of working horse collars.

Crupper – is a strap looped under the top of the tail and fastened to the back of the saddle to stop it slipping forwards.

Breeching – is a wide strap running round the pony's hindquarters and attached to the tops of the girth to stabilise the saddle.

Hefting – (habits) means the survival skills related to the terrain of a herd's grazing territory which are passed on from adult to calf, lamb etc. Essential for the survival of herbivores on high altitude pasture in the winter.

Atholl Brose whisky punch – 3oz oatmeal, 2 tbsp. liquid heather honey, whisky to make up to 1 quart and 1 pint of water. Make a thick paste with the oatmeal and water and stand for 30 mins. Pass through a fine strainer. Save the liquid and discard the oatmeal after pressing dry. Mix the liquid and honey and stir with a silver spoon until well blended. Pour into a bottle, add the whisky and cork well. Shake before using.

Shooting Season Dates

Grouse (England, Scotland & Wales)
    12th Aug–10th Dec.
Red deer stag (Scotland) 1st July–20th Oct.
Red deer hinds (Scotland) 21st Oct–15th Feb.

CHAPTER THIRTY-TWO

Stunt and Film Horses

The 1959 Hollywood epic, *Ben Hur*, in which two thousand horses took part is remembered particularly for its shockingly dramatic, nine minute long, chariot racing sequence. However, the high equine attrition rate during its production would be entirely unacceptable now as the welfare of performing animals is a priority.

Behind today's blockbuster films such as *War Horse*, live performances at Olympia and the Horse of the Year Show and television productions lies some very special horse work. On the screen, the old genre of Western films has been replaced by historical and futuristic dramas, such as *Victoria* and *Game of Thrones*, while in the arena there is a renewed repertoire of entertainment: Cossack acrobatics, haute école and also liberty work where there is no physical contact between the handler and his horse.

The training methods used for these displays have an affinity with the circus, the cavalry, police horses, natural horsemanship and dressage.

Atkinson Action Horses grew out of a livery yard on a Yorkshire dairy farm. In the twenty years since Mark Atkinson was first asked to provide and ride a horse for a historical re-enactment event, he and his son Ben have performed at top shows and taken part in many television series.

It is a family business: Jill, Mark's wife, is the co-ordinator and Lucy, Ben's sister, also rides in the team when she is not doing her job as a teacher. Ben's grandmother is a professional designer who, with seamstress Joyce Bentley, creates many of the costumes for both riders and horses.

Ben Atkinson, age twenty-four, is the performer and choreographer for the live displays and he and his father, Mark, work together on many television and film productions as horsemasters (supervising all the horse work on set).

Ben was well ahead of his peers having progressed quickly through the early phase of being a cowboy with cap guns blazing, galloping his Shetland to the horizon on a miniature Western saddle, to being a Roman standing on the backs of two ponies with a leg on each. Four generations of his family have farmed the land where he lives and where his grandfather taught him to drive and harness horses to both a plough and a carriage. And his father took him hunting – all before he was twelve. When Ben was fourteen, English Heritage sponsored him to learn the art of Cossack riding. He studied with Guido Louis and toured with his company, Rockin Horse, giving performances in Europe and the UK, and for classical dressage lessons, Ben went to Madrid.

It was working out his own way of communicating with horses that led ultimately to his present performances. His two main principles are: first, to gain the horse's complete trust and confidence in him and secondly, to build up the horse's fitness and strength to make it easy and safe for the horse to perform.

By watching experts on liberty work, such as Lorenzo and Jean-François Pignon, where the horse is free in the arena, Ben noted exactly where to stand and how to move his fingers, hands, arms and body in a way which meant something to the horse.

Ben Atkinson of Atkinson Action Horses, demonstrating mutual total trust during a manoeuvre in which his horse, Malik, is both unstable and vulnerable, yet relaxed.

By observing horses' natural herd behaviour and experimenting at home, he developed his rapport with them to a remarkable degree.

Working in an arena with a horse without saddle or bridle who was willing to 'join up' (to focus completely on him) he could signal to it to come towards him, go away from him, walk, trot, canter or stop round him. The signals are very subtle: spreading the fingers wide, for instance, is a sign to the horse to increase its pace. Further work would then be done with the horse attached on a long line, sometimes with a rider, until it understood what was required, before the line would be unclipped to continue the performance 'at liberty'. The whip carried by the horsemaster is used as an extension of his arm and never as a punishment.

Ben trains each horse over many weeks, working in very short sessions of less than half an hour, always making sure a manoeuvre is securely learnt before progressing. For example, in haute école to perform a levade in which the horse crouches on its hind legs and lifts its body to an angle of forty-five degrees, it must first learn the piaffe. The piaffe is a movement in which the horse's profile is shortened and rounded and its energy is directed to actively trotting on the spot in slow motion. The horse's centre of gravity gradually moves back to its hindquarters so that it can lift up to the levade position when the hind legs are well enough under its body to support its weight. Similarly, to safely ride a horse without a bridle, it must first learn to respond to advanced aids or signals from the rider's body and legs, without relying on the reins. Small shifts of weight from the rider are sufficient to indicate direction and pace to a trained bridle-less horse.

Several horses may be put together as a team once each has learnt the basics. 'This is sometimes easier than working with a single horse,' explains Ben. 'If I ask four horses to bow down together, I would put the two experienced ones at either end of the row and the two novices in the middle. There may be an imitative element where the two middle ones copy the outer ones but their natural pecking order is more important to be aware of. From watching the horses out in the field, I know which ones are dominant and which are subservient in the herd order. If, in the arena, a dominant horse bows down, those below it in the pecking order are likely to follow suit.'

Equine psychology is also used during filming, for example when scenes are shot from a helicopter flying low above the horse's head. During the making of *Poldark*, Aidan Turner (Ross Poldark) is required to gallop across moorland while being filmed from above. Aidan rides one of the Atkinson horses, Seamus, an Irish draught, who is unfazed by drones and helicopters but any new horses used would have an older, experienced companion horse alongside (though sometimes out of camera shot) to reassure them.

Horses' temperaments vary, just as humans do; some are cleverer than others, some become bored easily. Understanding the individual characters is fundamental for keeping the horses calm and enjoying their work. Providing plenty of variety in their work and minimal time in the loosebox is Ben's approach. One of his Andalusian classical dressage horses, Malik, can switch off from the extreme concentration required for the classical dressage 'airs above the ground' by doing some work with the harrow on the farm. Another, Sebastian, enjoys

A few of Atkinson Action Horses' saddles. From left to right: a jousting saddle copy, a Portuguese saddle, a Cossack trick riding saddle and a Universal Pattern army saddle.

pulling a carriage for exercise when he can trot for miles, idling mentally and relaxing.

I notice a parallel to this as Ben and I sit in the kitchen chatting over a mug of tea while Jill wades through paperwork next door: the calendar pinned up behind the door shows that not just every single day is accounted for but almost every hour is too. With such a hectic schedule Ben is very good at chilling out in between commanding intense focus during training sessions, discussing scripts and performing.

There are rules in his equine relationship: of course the horse must always respect Ben's personal space and concentrate on Ben despite distraction, but their observance is not through his domination of them. It is the result of mutual respect and training by reward – not with carrots but with praise so that the horse finds it is more comfortable and pleasurable to join in with the game. It appears to be an equine trait to be willing to please as a default attitude (unless spoilt by previous handling), as though the horse recognises a leader whom it can respect and trust and, having been convinced, is prepared to follow.

The tack rooms are full of saddlery from many historical periods and different countries. If a particular saddle is required, such as for jousting, the two saddlers on the team are skilled at 'building' a replica, using the Army UP (Universal Pattern) as the basic frame. Ben's favourite personal saddle is a

Ben Atkinson's Camargue saddle used for classical dressage displays.

Ben Atkinson with Arthur performing a spectacular capriole at the Horse of the Year Show in 2017 as part of the Atkinson Action Horses' display 'Limitless'

Wintec dressage saddle but for classical dressage he prefers a Camargue saddle. Trick riding saddles vary in design according to their use. Most have a triple girth and a metal clip from the girth to the stirrup for the rider to hold while hanging off to one side when the horse is galloping. The seat of the saddle is soft and padded to give grip to the rider's feet when standing and leather loops and horns are useful for vaulting and other acrobatics.

At the Horse of the Year show in 2017, Atkinson Action horses performed their show 'Limitless'. The thousands of spectators were buzzing with excitement following a tense jump-off in the event immediately preceding the Limitless display. Yet, as soon as Ben and Malik entered the arena, the mood changed to one of quiet anticipation – the concentration of Ben and his horse on their work was infectious. What followed was an impressive range of skills by the team – from acrobatics at the gallop to a dramatic capriole, jumping and liberty work – taken from several disciplines and united by perfect communication between individuals.

The Devil's Horsemen

The Devil's Horsemen in Buckinghamshire is the largest film and horse entertainment business in the UK. Its French founder, Gerard Naprous, arrived in Britain fifty years ago with little more than a background of stunt work learnt from a Czech cavalry officer, and a deep knowledge of the history of classical dressage in Europe.

He and his son, Daniel, and daughter, Camilla, organise equestrian performances on a grand scale and work with about eighty horses. Dan is also a successful international carriage driving competitor.

Gerard Naprous overseeing the building of a copy of a World War I ambulance for War Horse with a Roman chariot in the foreground.

A small sample of The Devil's Horsemen's tack from many different countries.

The French flag flies high over a complex of barns and workshops as big as a village and surrounded by fields. The level of activity is intense and hardly slackens off after dark; several carriages are being prepared for overnight transport to London for a film shoot the next morning just as two horseboxes arrive home from filming on location in Scotland. In one manège, a novice actor is having riding lessons and in another, a young horse is being carefully taught how to fall onto the sand without hurting itself. These highly skilled stunt horses occasionally show up any inexperience in their celebrity riders; a well-known actress riding side saddle during preliminary filming inadvertently gave the aid to fall while cantering. Obediently, the horse dropped its shoulder and fell, together with the actress who fortunately was quite unhurt.

Among the warren of workshops with lathes and industrial sewing machines we came across a horse-drawn carriage under construction. Like the horses, most of which can be ridden and driven, the workforce here is multi-talented and can improvise props for any set. Gerard's knowledge of historical detail is encyclopaedic and the style of horse and its saddlery or harness is correct down to the last buckle.

Gerard is adamant that his horses should be trained in 'dressage en basse école' which is basic dressage with rhythm in all paces, balance and suppleness. This is the equivalent of an athlete's fitness training. 'It is the basis for everything', he says, 'and as essential for the horse to joust safely as it is for falling at the canter. We use many European breeds of horse: Andalusian, Hungarian Nonius and Portuguese Lusitano here as they have the physique and temperament for this work.' He refers to the sympathetic training methods of Pluvinel, Louis XIII's Master of Horse who continued Xenophon's teachings from the 5th century B.C. despite the harsher methods practised in Italy in the intervening centuries.

A number of tack rooms here, including several lofts full of extraordinary treasures, hold around a thousand saddles. There is saddlery from the Napoleonic period, the American Civil war and pre-First World War Russian and Polish cavalry saddles with names and numbers stamped into the leather. Gerard brings history alive, talking throughout as we squeeze down an aisle between shelves loaded with richly decorated Gaucho and Barbarian saddles, mule pack saddles, English side saddles, children's donkey cart harness and more. He points to a Universal Pattern British army saddle explaining that it was based on a Hungarian design in the 1800s, and adds in passing that our English carriages were copied from the French in the 1600s and the Chinese invented the breast harness. We finished the day with Gerard driving a four-in-hand carriage round the farm. His vivid equestrian history lesson continued, drawing together the international strands of horsemanship with the same familiarity and panache as he handled the reins of the team in front of him.

Dolbadarn Film Horses

Dylan Jones, the Welsh horsemaster, lives in a Victorian coaching inn deep among the mountains of Snowdonia. I visited on a day when the valley floors and high peaks were hidden in mist and looked as dramatic as any film set. It was working with The Devil's Horsemen on the film *Merlin*, shot on location here, which began Dylan's career. He was already well prepared. He had been brought up working with horses in the family trekking business and having a fine singing voice had studied the performing arts at Bangor University.

Five years of riding with stunt team, The Horsemen of the Apocalypse and learning from different horsemasters, was followed by classical training in Portugal on Lusitano horses. After over twenty years in the business, Dylan is as busy as ever, training horses and riders and currently filming in London on *Aladdin* and in Scotland on *Mary Queen of Scots* with the Devil's Horsemen.

Dylan has continued to use Lusitano horses, mostly stallions 'They're clever enough to understand the most nuanced aids', he says. 'For instance, I'm often holding the reins in only one hand with a sword, lance or flag in the other. I can control the horse with the slightest body weight shift of pressure or turn of my shoulders. If forty of us are charging about at once, a horse that relies only on rein control would be a hazard. Also, they generally have calm temperaments and importantly, know how and when to relax. Not every horse is suitable though and each must be treated as an individual.' Most of Dylan's tack is handmade in Portugal but for dressage his favourite saddle is an elderly Albion English saddle.

Dylan also emphasises the importance of basse école; this forms the basis of a partnership in which the horse and rider learn to trust one another completely. The horse learns its paces and rhythm and to go forward without hesitation, knowing that it will be safe whether it is asked to take part in a cavalry charge or go through fire. Dylan's skill is in imparting reassurance to the horse and in assessing the individual's strengths and weaknesses. Building on its strengths first to give it confidence, he can then work on areas the horse finds more difficult. In this gradual progression his horses can develop seven or eight skills, such as carriage driving, trick and stunt work, liberty and classical dressage, combat or being a schoolmaster to actors who are novice riders, in contrast to most other horses who work in only one or two disciplines.

'When it comes to stunts such as falling at speed, not something I would do with my horses, there are rough and ready methods which achieve this by fear but you would then have an unreliable horse which could be dangerous and it would completely defeat the object of the training which is to maintain that mutual trust. My horses can bow and lie down because they have been taught gradually and gently so they have no fear and are willing partners', he continued.

All of the three horsemasters described here emphasise the trust which must be carefully established, an achievement which is proved in every performance by the willingness of their horses to present extraordinary feats.

Bartabas

A different form of entertainment, in which the horse is central to the performance, is emerging as equestrian theatre. Here the horses are ridden or are at liberty on the conventional stage in a traditional theatre, as seen at Sadler's Wells, London in 2011 and 2016, in a unique combination of equestrian technique and emotional expression.

The Frenchman Clément Marty, self-styled as Bartabas, founded Zingaro in 1984, a combined theatre and circus in Aubervilliers near Paris, before establishing the Versailles Academy of Equestrian Arts. Bartabas is a maverick horseman in the sense that he does not conform to any one discipline but is a master of many, and has an instinctive power of communication with Equus.

Only the best will do and he has chosen Hermès to make his saddlery. His film *Mazeppa* won the technical grand prix at the Cannes Film Festival in 1993 and he has been the recipient of numerous awards.

Bartabas' aim is to celebrate the horse by using it as a unifying theme in a fascinating mêlée of the art, philosophy, music and dance of many different cultures. His theatre is dramatic and thought-provoking: in Golgota (2016) the horses moved silently on the stage (covered in rubber chippings) as part of a religious narrative – though whether God, Man or the Horse was the true object of worship was open to interpretation. Once again, the horse has extended its remarkable repertoire.

Dylan Jones of Dolbadarn Film Horses. This Lusitano stallion, Valmorim, was bought as a foal and trained by Dylan to do a wide variety of stunts and trick riding.

CHAPTER THIRTY-THREE

Isle of Man Tram Horses

The Isle of Man has traditionally been famous for horse-drawn trams, kippers, Manx cats and the TT (Tourist Trophy) motorcycle races. Now in the 21st century, high-tech and film production companies are also thriving there. Its Tynwald parliament which dates back to Viking times, is the oldest continuous parliament in the world. The Island has its own flag, national anthem, bank notes and Manx Gaelic language and the Clydesdale and Shire horses which carry out public service work on the trams are bred and trained here.

Approaching on the Liverpool–Isle of Man ferry, passengers can see the welcoming curve of Douglas Bay with its elegant Victorian villas looking over the beach. There is light traffic along the main promenade but the horse-drawn trams plying to and fro at a spanking trot immediately catch the eye. The three-foot gauge tramway, acknowledged to be the oldest operating horse-drawn tramway in the world, was built in 1876 by Thomas Lightfoot, a Sheffield engineer who retired to the Isle of Man. The trams, which run for two miles between the ferry terminal and Derby Castle near the tramway stables, are popular with the tourists and have outlasted two world wars and the introduction of buses and cars.

Each tram, seating some twenty to thirty passengers, is pulled along rails by one horse. At the ends of the route, the horse is unhitched, led around to the rear of the tram by the driver and conductor and hitched up again in a matter of minutes. On one rare occasion when a horse called Duke (after the TT race champion Geoff Duke) was frightened by a loud bang during a changeover and escaped, he galloped back to the stables keeping between the tram lines all the way and arrived safely.

The trams, all in red and cream livery, are of different designs and kept in good condition by a refurbishment programme. Each can weigh about two tons. It may seem a heavy load, especially when watching a horse lean into his collar to start the wheels rolling after a tram stop, but the rails enable the tram to travel more freely and the axles have roller bearings for easy movement. As with pulling a canal barge, starting off is a knack which the horses soon learn. After that the momentum of the vehicle makes light work. Indeed in 1969, a visiting celebrity famously put a horse collar round his chest and demonstrated how easily he could pull a tram with twenty passengers by himself.

A strong spring connected to the traces absorbs any shocks and the horse's heavy-duty shoes have studs to grip the special surface between the rails. Interestingly, the area of wear on the shoes is usually greatest towards the heels, quite opposite to a riding horse which wears its shoes at the toe. This pattern of wear is consistent with the horse's draught power coming from its hind legs as it pushes into its collar to move a load, rather than simply pulling.

As the trams are equipped with brakes and the route is level, minimal harness with an English collar and traces, without breeching, is worn. Some horses are happy in blinkers, others prefer an open bridle and many wear snaffle bits.

The Douglas Bay Horse Tram Company has recently been amalgamated with the Steam and Electric Railway companies by the Tynwald

THE TACK ROOM

Isle of Man tram horse 'John' plying along the seafront between Victoria Pier and Derby Castle

government and provides an excellent lifelong career for these horses with a secure retirement.

Heavy horses are bred for strength and have a good temperament for draught work. During their training for eventual tram work, the horses are long-reined on the beach to reach the level of fitness required for the tram route.

The salt water is good for their feet but the exercise serves another useful purpose. The response of the horses to the breaking waves is a gauge of their temperament and suitability for road work; the tram lines run between two lanes of traffic and a steady horse is essential.

Most horses are not upset by the waves but the occasional individual finds them too alarming – perhaps, one could imagine, spooked by Manannan, the mythical sea god, whose island kingdom here was known to Neolithic, Celtic and Viking inhabitants.

The horses work shifts of approximately two hours a day. The tram service which at peak times runs at ten minute intervals, requires four trams and if each horse works a two hour shift (of about ten miles), five horses per tram and a total of about twenty horses are needed.

A fine tram horse waiting for his harness to be removed before being washed down after his two hour shift. The tractor is used to clear out the muck heap.

Some of the 21 Clydesdale and Shire tram horses ('trammers') in their stables on Summerhill Road.

In the 1880s in the heyday of Manx tourism, the trams carried over six thousand passengers a day. In the 1920s, buses were given permission to travel on the promenade during the winter storm season when waves would sometimes flood the road, and the horse-drawn trams operated in the summer.

Members of the public are welcome to visit the stables and the friendly atmosphere is reflected in the bright and willing attitude of the horses. The working life of these horses varies but can be about fifteen years after which they remain on the Island in retirement.

The horses are stabled in 19th century stone barns where their looseboxes have low partitions for easy socialising between each other. The barns are well lit and comfortable and full of the scent of first class haylage.

Michael Crellin, manager, has been with the horses since he was an apprentice smith at the stables in the 1970s, when each shoe was made to measure from a length of iron. Now imported Dutch blanks are used.

As we walked to the palatial harness room, which he has designed, he told one of the many legendary stories about the horses. One of the Clydesdales kept stopping in front of a particular hotel on its route. Eventually it was explained by the fact that the horse, who recognised the conductor's whistle signal to stop, was indeed obeying the whistle – but it came from a parrot in the hotel who had learnt to imitate it, and not from the tram conductor.

The harness room occupies the full extent of a loft in one of the barns. It is warm and bright, top lit with plenty of space for the harness as well as desks for office work. In the centre is a large table, topped in impervious stone with raised edges and an integral sink, which is ideal for cleaning tack. It is the best design of table for the purpose I have seen anywhere.

At the end of World War II when the tram service re-started, 42 horses of mixed types, known as 'vanners' were imported from Ireland. Their arrival, trotting along the promenade, caused the residents to think the cavalry had arrived. Although they were more economical in the wear and tear of their shoes, it was decided in the mid-1970s that it would be cheaper to breed Clydesdales and Shires for tram work on the Island. As these were much bigger than the vanners, new and bigger sets of top quality harness were made in Walsall. Each of these collars, with solid steel hames and the horse's name on a leather plate, hangs above its bridle round the walls.

The superb harness room at the Douglas Bay Horse Tram Company stables. The set of black collars on the beam were for the original vanner horses (mixed draught breed) from Ireland which are too small for the island-bred heavy horses used for the tram service now. The new harness and bridles with blinkers, made in Walsall, are arranged on red racks round the wall.

An electric light trap for any moths which might attack the woollen collar linings and a resident cat to deter mice are important to protect the harness from pest damage.

In past times, the tram horses were loaded onto the ferry for slaughter on the mainland after their public service had ended. One day in the late 1940s, Mrs Mildred Royston and her sister, Miss Kermode, noticed the horses slipping and falling on the wet ferry ramp as they were being loaded. They demanded that straw should be spread to help the horses.

Hearing that this was to be their final journey after years of service, they eventually persuaded the authorities to discontinue the horse export. Determined to give these working horses a dignified retirement they established a Home of Rest for Old Horses in 1950 which is now at Bulrhenny in the lovely countryside outside Douglas. This unique life for a working horse, born on the Island, providing a public service and enjoying guaranteed retirement, represents a privileged equine existence.

CHAPTER THIRTY-FOUR

Western Saddlery

You can find Western saddlery in a surprising number of British tack rooms. It is of European origin with its own special history of development dating back to 1519, the arrival of the conquistadores on the Mexican coast. Such an exact date for the beginning of equestrianism in a continent is unique in the equine world.

There had not been any horses in the Americas for thousands of years before 1519. Christopher Columbus had made the first landfall in the Dominican Republic on his second expedition in 1494 with five mares but their survival is doubtful. Horses were driven overboard to swim to shore or were slung onto small boats but, on reaching land, the jungle conditions were far from ideal for horse or man. Hernán Cortés (1485-1547) is credited with landing the first surviving horses in 1519 during his campaign to subdue the Aztecs in Mexico.

Ten thousand years before Columbus arrived in the New World, there had been a mass extinction of the megafauna. The primitive equids, which were the ancestors of today's domestic horse, died out leaving a continent without horses. The cause of this extinction is unknown but climate change or hunting by humans has been suggested. It is widely accepted from fossil evidence that horses had originated in North America and, before the extinction there, some had spread across the Bering Straits land bridge to the rest of the world.

The remote ancestor of the modern horse (*Equus ferus caballus*), known as the 'dawn horse' (*Eohippus*), had toes instead of hoofs and was the size of a dog. Over tens of millions of years equids became larger and their toes developed into a single hoof. Those that reached the Asian steppes flourished on the grazing and diversified as onagers and tarpans (now extinct), asses and horses while others migrated into Africa and evolved to become zebras.

After this extraordinarily long interval in which the horse was absent in the Americas, it is remarkable how quickly it thrived following its reintroduction in 1519. Descendants of the conquistadores who landed in Central America dispersed northwards to become ranchers, taking their horses with them.

By the late 1600s horses had reached the grazing lands of the Southern States either with early colonists or in feral herds and a hundred years later they had populated the mid-west up to the Canadian border.

The waves of settlers from the 16th century onwards brought many contributions to American life, one of which was the design of its saddlery. Today's typical Western styles gradually developed from the Iberian war saddles of the Spanish conquistadores. A statue of Cortés on his warhorse El Morzillo shows him and the horse in full armour. The straight leg and long stirrup style of riding (known as a la estradiota), necessitated by the armour plate, required a turret-like saddle with thigh supports, a protective pommel and a high cantle for the rider to push back into. As guns replaced swords and lances there was less need for full armour. Lighter, faster cavalry with flatter saddles and shorter stirrups became more effective on campaigns.

The Mexican vaquero (cowboy) saddle was taking shape by the 1700s by expanding the Spanish

A bespoke Western show saddle built by Kevon Trusselle, Wolverhampton, UK.

By the 1860s the carved wooden stirrups made from single blocks of wood were replaced by lighter woods such as the ponderosa pine or cottonwood, steamed and bent into shape and often covered in leather. Wood was better than metal for stirrups in extreme temperatures, lighter and more available. A leather hood or tapadero sometimes covered the front of the stirrup to keep the foot warm and protect it from thorny brush.

Metal horns of steel, brass and nickel, bolted onto the wooden saddle tree were common by 1900 and over the years regional saddle styles developed such as the Californian with rounded skirts and the Texan with squared skirts.

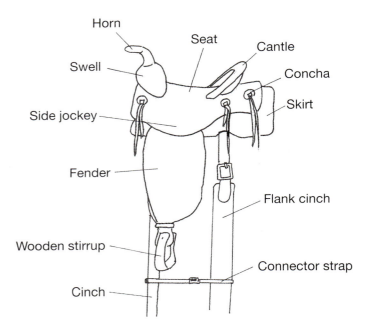

The Western saddle with horn and two cinches for roping.

saddle seat to make it more suitable for working with stock. The tree was made up of wooden components comprising a frame and covered in rawhide. The weight-bearing bars extended backwards on either side, the cantle and pommel were lowered and initially there was no horn.

As colonisation continued and cattle ranching expanded, the saddle acquired a horn in the 1820s. To counteract the force of a lariat tied to the horn under tension 'double rigging' (two cinches or girths) was introduced to prevent the saddle from tipping up.

The rear cinch, loosely fitted round the horse's flanks was connected to the front cinch by a strap under the belly which stopped it swinging too far back and damaging or frightening the horse.

A removable leather seat cover (a mochila), through which horn and cantle protruded, covered the tree and was widely used. The 1860 Pony Express mochilas had lockable mail pouches on each corner and could be quickly transferred to a fresh horse's saddle at each station.

The westward tide of colonists, swelled by the Gold Rush in 1849, initiated the Indian Wars in the 1860s in which another mainstream type of American saddle was important. The Hope saddle used by the US cavalry in the 1850s was of a simple English style (which was based on the Hungarian Hussar saddle) with a small horn. It was followed by George B. McClellan's design of 1859 which is far better known. The modern equivalent of the McClellan saddle is still the preferred choice of some trail and endurance riders today. It has no horn and a medium height cantle and pommel. The bars of the tree extend behind the cantle to which a pad can be buckled to carry light provisions. It is a minimal saddle (early versions had no skirts or jockeys) which sits on a folded woollen blanket for a pad and is

Right: This Native American saddle was recovered from the battlefield of the Little Bighorn, where Custer's Last Stand took place in 1876. The broken hook on the high pommel would have been used to carry an infant in its papoose. It is known that Lakota and Cheyenne women took part in this battle. The saddle was from the late Charlie Clayton's estate and was given to him in 1960 by Red Woman whose ancestors found it in the Black Hills.
Photo courtesy of Sally Mitchell, Museum of the Horse, Tuxford.

Below: A Native American beaded bridle with jointed metal bit and American eagle feather decoration. Date unknown. Gifted to the late Charlie Clayton. Photo courtesy of Sally Mitchell, Museum of the Horse, Tuxford.

Hand tooling of a typical botanical design on a custom-built Western saddle. The design of stylised native wildflowers is accompanied by foliage often resembling the leaves of Acanthus spinosus, which can be elegantly adapted to fit any space on the saddle.

secured by straps from the pommel and cantle which connect to a single central ring through which the cinch is fastened (centrefire rigged).

The Native Americans who had adopted the horse into their culture rode bareback but both men and women also used saddles. Horses were chiefly used by the buffalo-hunting peoples on the Great Plains and in the Dakotas though even they had not assimilated the horse for more than a few generations. Paintings by Frederic Remington show that Native American braves used stirrups and made their saddles with high pommels and cantles.

The leathers used in Western saddles are more varied than in English saddlery. Latigo leather, introduced by the Spanish is tanned with a combination of alum and vegetable tanning and is strong and flexible enough to attach the cinches to the saddle. The best rawhide (incompletely tanned leather) is bull hide which is used to cover the horn, tree and stirrups and to form the noseband of a bosal, a bitless bridle. Wet rawhide shrinks as it dries and forms a tough reinforcing skin.

The saddle was as extravagantly decorated as the owner could afford and it would have been his most precious possession. It is thought that patterns on the leather were originally made by cowboys heating a nail or spur in the campfire and scorching a simple design on the saddle such as a basket-weave. Traditional patterns tooled on Western saddles are botanical with native flora, though the acanthus leaf and rose design by William Morris, fashionable in England in the 1870s, has also been used. Bespoke saddles can be art forms and heirlooms, such as one commissioned by a John Wayne fan, which is covered in detailed scenes from each of his films, and another in which jade and semi-precious stones are incorporated into the leather. Both are in private collections. The surface of such tooled saddles provides grip as well as some ventilation in the heat.

Al Stohlman (1919-1998) was a Californian who pioneered ornamental leather work and introduced new tools and methods which popularised saddle decorations. He and the Tandy Leather Co., which continues to supply tools, leather and books, had a strong influence on Western saddlery ornamentation. In the UK, Brian Borrer and Kevon Trusselle are expert Western saddle makers, fitters and braiders.

Western saddles today have trees of wood, injection-moulded polymer or fibreglass and can be highly decorated with silver conchas (screws with ornamental heads holding layers of the saddle together) and elaborate tooling. The mochila is rare now and leather seats are fixed. The leather skirt under the length of the seat is usually lined with sheepskin and sits on a thick saddle blanket over the horse's back.

Comfort for the cowboy lies in a well-designed seat. For example, to avoid leg chafing, the stirrup leather usually lies under, and is attached to, a broad

A corner of the tack room at Sovereign Quarter Horses.

piece of leather called the fender which also keeps the horse's sweat off the rider's leg.

Cinches may be made of leather, webbing, neoprene or of mohair in a 'string' design which is soft and cool. The saddle has numerous D-rings and saddle strings for the attachment of, for example, a bedroll and rifle scabbard.

There are variations of the Western stock saddles to suit different purposes: heavy roping saddles with strong horns, reining saddles with stirrups which swing forwards for the sliding halt, decorative parade saddles for showing and wide swells either side of the horn on cutting saddles to support the rider's thighs during sudden fast stops and turns when targeting an individual steer.

The bridle design is minimal, in that there are no nosebands, often no throat lash and a simple loop over one ear instead of a headpiece and browband. The cheek straps can be decorated with beading or silver and curb bits are commonly used.

Two Western designs of rein are the split and romal reins. Split reins are separate (not buckled together) and romal reins are a continuous loop from which there is a single rein extension of about three feet (this extension is the romal which can be used as a quirt or whip).

The reins are of heavyweight leather such as plaited rawhide, so that they drape and provide a balance between the hand and bit to ease the pressure on the horse's mouth. This delicate contact can be demonstrated by showmen who attach a horse's tail hair between the bit and reins on either side and show that the horse can be controlled without breaking the hair. This clearly illustrates the principle that it is how the tack is used (and adjusted) that is crucial; in unskilled hands the Western bits could be damaging.

The bits are reminiscent of 16th century European bits and are still used, particularly for reining, but the ports, spades (extra projections on the mouthpiece of the bit) and long shanks have to be within strict limitations for the show ring. Matthew Harvey Ltd. of Walsall was manufacturing such bits throughout the 19th century for the South American market. Silver ornamented bridles and silver trimmed saddles continue to be popular.

The method of training the horse to accept these bits is quite unlike anything in British schooling. This is achieved by using a bitless training bridle, a bosal, in which the rawhide noseband is attached to a heavy horsehair rope (a mecate) knotted under the horse's jaw close to the curb groove. The rein may be wrapped around the knot so as to alter its weight and counterbalance the noseband when the horse's head is in an acceptable position. If the nose is raised, the pressure from the noseband is increased, if the horse overbends to evade it, the knot under the jawbone discourages the evasion. When the horse maintains the correct head position, neither the noseband nor the knot should exert pressure on the horse.

Lighter bosals are used as the young horse learns to balance itself at which point it can make the transition to wearing the bit. Only the slightest touch on the rein, perhaps lightly weighted with decorative metal bands, is then needed for control. The bristly mecate rein is also used for teaching a horse to neck rein, encouraging it to turn away from the prickling against its skin. Neck reining is a major aid, leaving the rider's hand free for roping.

The Western style of riding, although not mainstream in the UK, has an enthusiastic following here and is regulated through the Western Equestrian Society (formed in 1985 and open to all breeds). The American Quarter Horse Association (AQHA) is the largest breed registry in the world (about three million horses).

The Quarter Horse descent can be traced through the mustangs and Chickasaw horses which had been developed by Native Americans from the Spanish conquistador horses with their Persian and Arab origins. Thoroughbred blood was introduced by English settlers. One of the most famous sires was Janus, a grandson of the Godolphin Arabian who was imported to Virginia in 1756.

Quarter horses are extremely versatile and famous for their sprinting ability, taking their name from being the fastest breed over a quarter mile race track. Their strength and agility became indispensable for working cattle and these attributes are tested in national and international competitions such as Barrel Racing, Horsemanship, Pleasure Riding, Trail, Ranch, Reining and Western Riding which includes elements of dressage such as flying changes.

Trail classes require horses to negotiate a number of obstacles including reversing back along an angled course and opening and shutting a gate. These manoeuvres would be essential for a stockman to close the gate on a corral of cattle without dismounting. Their origins lie in the long cattle drives where cowboys would buy herds of cattle in the Southern States for $5 a head and drive them up to the north-east to sell for $25 a head.

Reining is a test of balance and speed demonstrated by the execution of precise patterns, spins and sliding stops. It is a recognised discipline in the World Equestrian Games and Para-Reining is being considered by the FEI for inclusion.

David and Sarah Deptford of Sovereign Quarter Horses in March, Cambridgeshire have a well-established Western riding centre for breeding, training and competing. In 2015 they produced the European Junior Ranch Riding Champion and were finalists in 2017.

In their large indoor school complex are offices, a tack shop and a tack room packed with treasures. On entering, the colourfulness and exuberance of decoration on the tack was striking, quite a contrast to the understated British tack room. The bright geometric designs of a stack of Navajo blankets, braided reins and nosebands and the intricately patterned saddles showed the pride in craftsmanship and turnout. The competitive rider's outfit from neckerchief to boots and spurs is eye-catchingly vivid and often complements the horse's tack in colour and design.

David Deptford demonstrating sliding during a reining competition.

Conclusion

When I started this journey I intended to establish the tack room as a palace of sorts in its own right. By looking at its treasures I hoped to strengthen the links between those who make saddlery and those who buy it, cherish its craftsmanship and celebrate the great versatility of the horse in work and leisure.

I was looking for a veneration of tradition and the excitement of innovation and I found it in both quality and quantity. From some of the finest harness in the world at the Royal Mews to the cutting edge of design and manufacture in Walsall, the search exceeded my expectations. The tack rooms I visited varied from the dormant and dilapidated to the most modern and well-appointed. I found many unexpected gems and was privileged to meet people whose expertise and dedication were inspirational.

A wealth of craftsmanship, often taken for granted, has been highlighted and its evolution continues through the collaboration of riders, saddlers and apprentices, manufacturers and the veterinary profession. Top quality specialist leathers are produced by British companies and the Abbey foundry in Walsall manufactures lorinery for the home and export markets. The saddlers' apprenticeships supported by the Livery Companies are oversubscribed. Their traditional methods of working with leather are still superior and can be adapted to modern materials, such as a handstitched saddle with a carbon fibre or titanium tree.

This book has come too late for many tack rooms. Understandably, but regrettably, progress has pushed them aside. Economic necessity has no room for sentimentality and tack rooms can only exist as long as their owners continue to have an equestrian interest. All the same, it is hard to see the grand old places laid low; modernisation can sometimes look as though the period architecture is wearing an uncomfortable shoe.

I was passing Holkham Hall in Norfolk and found the stable courtyards cordoned off with red and white tape. Major building works were in progress to convert the stables into a commercial wedding venue but with the help of a sympathetic builder I went in search of the tack room.

Two men were stripping out the last wooden saddle racks, the wood panelling had already gone and nothing was left. It had been a large and splendid saddle room which now looked distressingly abandoned; all that history unrecorded and erased for ever. In the adjoining rows of stalls, also awaiting the same fate, were large mangers made of Welsh slate, an unusual alternative to cast iron. Transporting the slate from North Wales to Norfolk must have been no easy task but they made excellent mangers, more hygienic than wooden ones and I hope they fought valiantly against being dismantled.

Hovingham Hall in North Yorkshire has a fine early 'riding house' which is the entrance to the main house. The owner so loved his horses as to want them under his own roof – much to the dismay of the ladies in his family who complained about the smell wafting into their boudoirs. The riding house remains but the stables and the tack room disappeared long ago.

Castle Howard has a beautiful stable courtyard which has been converted for the comforts of thousands of visitors. Its tack room is now a shop, though still retains the original wood panelling, and the adjoining stalls are used for storage.

The magnificent stable yard of Althorp House, built in 1733 of local ironstone has been converted to include a café and conveniences, as at Cliveden and elsewhere.

In my childhood, tractors stormed into the working horse world and, particularly in East Anglia, heavy horses disappeared within weeks; their harness and horse brasses were shovelled into pits and the tack rooms were converted to garages. One of my hopes had been to find a tack room with a Musgrave saddle horse but, in all my travels, the closest was one of similar design but without the folding top, at Milton Hall. It shows how rare some of the fine old stable fittings have become.

Looking forwards though, new tack rooms are being built and it is the harness and saddlery within them that is important. They also continue to provide a focus for people to meet, learn and enjoy their horsemanship.

Some, such as the harness racing, carriage driving and some horse logging tack rooms, now differ fundamentally from their predecessors: the characteristic 'saddle soap and leather' smell has vanished. Synthetic leather, such as biothane, has almost taken over from leather for harness because it is practical and economical. In other tack rooms, show harness of leather decorated with shining metal work is still cherished and reminds us of its traditional magnificence. The riding world has largely retained leather saddlery but synthetic materials are establishing themselves in some endurance and racing tack rooms.

Tack and equipment has evolved with the times, often in response to the demands of more competitive professional riders. The development of design and manufacture is not only the preserve of older craftsmen but of a whole new innovative generation. New materials used in the manufacture of tack, such as weatherproof and cheap nylon webbing and carbon fibre, plastic and titanium used in saddle trees for weight reduction and strength, are all finding a place in modern tack.

In general, a more sympathetic approach to horses is evident today than in the past and the horse's comfort is increasingly considered in the design of saddlery and harness. More emphasis is placed on the correct fitting of saddles so that both horse and rider are comfortable, and qualifications for saddle fitters are regulated by the Society of Master Saddlers.

In the 21st century great progress has been made in the technology of pressure mapping where we are embarking on a whole new approach to the design of saddlery and harness. This is a significant leap forward in the history of saddle and harness-making (including bit ergonomics), particularly in their humane design.

Sensors on computerised pads can show localised areas of pressure and give us insight at last into what the horse may be experiencing as it works. The powerful photographic technique of gait analysis can give objective evidence for this in the measurement of freedom of stride and flexion of joints. Anecdotal evidence from owners confirms that better-fitting tack results in more compliant behaviour and better performances from their horses. Other technologies such as the application of medical thermography, which can identify inflammation and capillary blood flow changes below the surface of the skin, may also enlighten us further in the future.

Smart wear for horses is being strongly marketed. A girth developed in France (by Seaver) measures several parameters including heart rate and jump patterns, using Bluetooth technology. Headcollars, surcingles and rugs with sensors for monitoring behaviour and temperature can alert the owner's mobile phone to the possibility of colic or imminent foaling.

How we use the tack to control the horse is also increasingly scrutinised. New rules endorsed by the International Federation for Equestrian Sport (FEI), the British Horse Society, the British Racing Authority and other regulatory organisations forbid the use of punitive spurs, draw reins for rollkur, over-tight check reins and curb chains and excessive use of the whip in racing.

This progress may be due to the reach of the British Horse Society, Pony Club and various riding clubs and to the influence of inspirational teachers and role models such as Monty Roberts, Mark Rashid and Carl Hester.

This collection of essays is not a comprehensive manual of instruction but a keyhole glimpse into the many ways we work with horses. Any errors are mine alone and are not intended to offend or mislead. The hardest part has been to decide what to leave out and I am aware that I have only covered the tip of the iceberg.

In today's mechanised and electronic world horses are flourishing and are significant in many ways: economic, therapeutic and internationally competitive. Many rely on the horse to maintain a healthy work/life balance and one of its best attributes is that it continues to connect us with landscape and nature through many disciplines such as endurance riding, hunting, stalking, horse logging, long distance riding, farming and of course hacking round our local countryside.

I hope readers will enjoy this canter through time and place, looking forwards as well as backwards, meeting people who share a passion for the horse in all its glory and be encouraged to explore further.

THE HORSE by Ronald Duncan

This cavalcade of grace now stands,
It speaks in silence
Its story is the story of this land.
Where in this wide world can man find,
Nobility without pride, friendship without envy
Or beauty without vanity?
Here, where grace is laced with muscle,
and strength by gentleness confined,
He serves without servility;
he has fought without enmity.
There is nothing so powerful,
Nothing less violent; there is nothing so quick,
Nothing more patient.
England's past has been borne on his back.
All our history is in his industry:
We are his heirs, he our inheritance,
He is of course, the Horse.

This poem was commissioned by Mike Ansell in 1954 and is traditionally read at the finale of the Horse of the Year Show in London. (© Ronald Duncan Estate)

ACKNOWLEDGEMENTS

One of the special pleasures of this project has been the kindness of colleagues in introducing me to their friends who shared our enthusiasm. I defer to the expertise of many from whose encyclopaedic knowledge I have benefitted: particularly to Richard and Angela Gifford, Nigel Lithgow, Mary Tuckett, Nigel à Brassard, Toby Brown, Jane Lake, Roger Philpot, Mark Roberts, Scherie Dermody, Patricia Nassau-Williams and Master Saddlers Mark Romain, Frances Roche, Nick Creaton, Terry Davis and Laura Dempsey, Issi Russell and Tiffany Parkinson. Past Masters David Snowden, Harriet Coates and Martin Payne have all been most helpful.

Thank you to those who helped very significantly: Brian Borrer, Virginia and Michael Cunningham, Lisa Etheridge, Vanessa Fairfax, Megan Green, Roanne Evans, Andrew Harman, Diana Hommel, Kate Mobbs Morgan, Sheelagh Neuling, Jim and Jean Power, Steffi Schaffler, Ian Stark, Hugh Toler, Mary Tuckett, Elaine Walker, Minna and Peter Wigley and General Sir Evelyn Webb-Carter.

I greatly appreciate the time taken by all those who generously allowed me to visit their tack rooms or workshops: Ben Atkinson, Andrew and Annalisa Balding, the Duke of Beaufort, Paul Belton, Emma Bletcher, George Bowman, Shane Breen, Philip Brind, Gregory Brown, Debbie Burton, Michael Crellin, Graeme Cumming, David Deptford, Boyd Exell, Richard Farrow, Nell Gifford, Rylla Gurdon, John Henderson, Liz Henderson, Nicky Henderson, Carl Hester, Brian Higham, Household Cavalry Mounted Regiment, Tricia Hurle, Jim Johnstone, Dylan Jones, Sister Mary Joy Langdon, Jeanette Keeley, Mary King, King's Troop Royal Artillery, Merseyside Mounted Police, Jim McInally, Lynne Munro, Gerard Naprous, Shirley Oultram, Philip Naylor-Leyland, Michael Owen, the Lord Palmer, Andrew Parr, David Poxon, William Reddaway, Annie Rose, Philip Ryder-Davies, Urs Schwarzenbach, Arnold Smith, Mandy Stanley, Sorrel Taylor, Nicki Thorne, Terry Vincent, Richard Warburton, Jonathan Waterer, June Whitaker and Helen Yeadon.

Several museums have provided specialised information, particularly Sally Mitchell's Museum of the Horse, Tuxford; the National Horse Racing Museum, Newmarket; the National Collection of Leather, Northampton and the Leather Museum, Walsall. These are invaluable resources for the general and equestrian public. Source material from the written legacies of Aimé Tschiffely, Lady Wentworth, Elwyn Hartley Edwards, J.W. Waterer, Giles Worsley and George Ewart Evans was most valued.

I owe a debt of gratitude to all the photographers. My special thanks go to Pamla Toler who endured thousands of miles in the car visiting tack rooms with me, not to mention the top of the Grampians in a snowstorm and the depths of Big Pit, the South Wales coalmine. I am very grateful to those who also gave their time free of charge to the project and for World Horse Welfare; Margaret Salisbury rose expertly to the challenge of aerial photography and Lynne Shore and Max Wenger went out of their way to help, as did Rod Blackmore, A.J. Booth, Roger Harris, Sian Davies, Paul Lunnon, John Minoprio, Nico Morgan, Roy Peckham, Giles Penfound, Andrew Rees, Gem Hall and Richard Stanton.

I am very fortunate indeed to have been given an adventurous brief by the publisher, Merlin Unwin Ltd. The tact, humour and excellent advice from all the team have kept me within generous boundaries of this infinite subject so that the making of this book has been a huge pleasure throughout.

Lastly, I thank my family: my sons, Edward, for organising my IT system to create order out of chaos and Patrick, for his expert advice on veterinary aspects and equine welfare; my sister Fiona, for her equestrian skills and piloting the plane for aerial photography and my husband Robert, for manuscript reading, despite being allergic to horses, and for holding the fort at home during my many absences. Thank you to my late mother, Ben Muir, for setting my life's compass towards the horse.

GLOSSARY

Aftwale: The leather-covered padded area of an English draught horse collar.

Breeching: A wide strap passing around the horse's hindquarters used for braking or reversing a vehicle.

Brash: Twigs and small branches left on the forest floor after the main branches have been lopped off (snedded) from the main trunk.

Cart saddle: A well-padded wooden frame placed on the horse's back which connects the harness together. The collar is strapped to the front of it by meeter straps. The crupper strap is attached to the back of it and loops round the tail. In the centre of the wood frame is a transverse channel for carrying the ridger chain which is attached to and supports the shafts. The cart saddle is secured with a girth.

Collarcheck: Traditional woollen cloth used to line working collars and cart saddles or pads.

Crupper strap: a stabilising strap which runs along a horse's back from the rear of a riding or cart saddle ending in a loop (crupper) through which the tail is threaded.

Feathers: The characteristic long hair on heavy horses' legs below the knee and hock.

Four in hand: A team of two pairs of horses, one pair behind the other.

Forewale: Hard leather roll filled with rye straw at the front of an English draught horse collar.

Hackamore: A bitless bridle which can exert pressure on the nose and poll of a horse instead of the mouth.

Hames: Metal or wood lengths which fit into the forewale of horse's collar and are fixed top and bottom by straps or chains. They provide rigidity and points of attachment for the traces or tug chains and terrets (rings) for the reins to pass through.

Hands: Units of measurement of a horse's height which is the length of a vertical line drawn from the withers to the ground. A 'hand' is 4" (10cm), originally the width of a man's palm. Thus 18hh (hands high) is a height of 72". The 'withers' is the bony projection of the horse's spine where the neck rises from the back.

Hefting: (as in grazing habit) survival skills related to the terrain of the grazing territory.

Horse breeds (heavy): Ardennes, Belgian, Brabant, Clydesdale, Comtois, Percheron, Shire, Suffolk Punch.
Horse breeds (light): Andalusian, Appaloosa, Arabian, Cleveland Bay, Dales, Eriskay, Galloway, Highland, Irish Draught, Lipizzaner, Lusitano, Nonius, Quarter horse, Shetland, Standardbred, Thoroughbred, Welsh Mountain, Welsh Cob, Windsor Greys

In hand: Leading a horse rather than riding it.

KWPN: A type of horse registered with the Royal Warmblood Studbook of the Netherlands (KWPN).

Landing: An area where the extracted logs are stacked awaiting collection by lorry, usually alongside a road, hard track or on a hard surfaced area (hard standing).

Leader: The horse harnessed in front of the wheeler horse which is hitched to the carriage or cart.

Lines: A ploughing term, also called long reins and made of rope or leather about twelve to fourteen feet long. The lines are connected to the horse's bit, threaded through rings (terrets) on the harness and held by the handler who guides the horse from behind.

Liverpool bit: A traditional bit used for coaching and draught work. It has a straight bar, sometimes with a port, through the mouth and a long shank either side. On the shank are two or three attachment rings for the reins, each providing a different degree of leverage.

Long gears: A simple harness used by a horse to pull a log or bundle of saplings along the ground, also called snigging or tushing. Chains are hooked to the horse's collar, and extend backwards on each side through supportive loops on the harness to a swingle tree to which the load is attached.

Oxer: A show jump with parallel bars.

Poll: Area on the top of a horse's neck immediately behind its ears.

Rack: A route for extraction formed by cutting out a line of trees in a planted stand of trees.

Saddle Makers: Albion, Amerigos, Barnsby, Bruno Delgrange, Camargue, Champion & Wilton, Fairfax, Free n'Easy, GFS, Hermès, La Martina, McLellan, Mayhew, Ortho-Flex, Owen, Pathfinder, PDS, Reactor Panel, Saddle Exchange, Stride Free, Toptani, Wintec.

Swingle tree: A wooden or metal bar fixed between the draught chains or traces at the back of a horse to keep them away from its sides and to which a load may be attached. Also called a whipple tree.

Tree (as in saddle tree): The internal skeleton of a saddle. It is a rigid structure of wood, metal or synthetic material around which the saddle is built, designed to protect the horse's spine and withers from pressure.

Thinnings: Young trees which are felled and removed to allow better quality trees to thrive. Second thinnings may be taken out at a later time, depending on the species of tree and the rotation time (planting to felling) of the crop.

Wheeler: The horse harnessed nearest to the vehicle.

Whipple tree: *see* Swingle tree

Woodflour: Very fine sawdust used to whiten and enhance the horse's feathers.

Also published by Merlin Unwin Books www.merlinunwin.co.uk

The Byerley Turk
The Ride of my Life
Horse Games
Right Royal

Horse Racing Terms
The Racingman's Bedside Book
Saddletramp
Vagabond

INDEX

Abbey England Ltd 15, 35, 46, 47
Albion Saddlemakers Co Ltd 9, 10, 46, 195, 227
American barn 124, 185, 211
Amish 115, 116, 146, 147
Animal Health Trust Centre for Equine Studies 19
apprenticeships 11-17, 30, 148, 204, 239
Arlington Court xi, 82, 83
army 72-79
Astley, Philip 89
Atholl Estate 215-220
Atkinson Action Horses 221-228
Atkinson, Ben 221-228
Badminton House 108, 130, 131
Badminton Horse Trials 5, 108, 109, 110, 112, 130, 191
Baker & Co Ltd, J & FJ ii, 2-4, 35, 63, 75
Balding, Andrew 184-185
Balding, Clare 174, 177
Balding, Ian 184
Bartabas 227-228
Beaufort, 10th Duke of 108, 130, 131
Beaufort Hunt 127-139, 138
Belton, Paul 9, 195
bits 41-7, 51, 53-4, 59, 64, 70, 72, 75-6, 82-3, 85-6, 94, 96, 98, 100, 105, 111, 116, 122, 134, 171, 173, 178, 183, 190-1, 194, 204, 207, 218, 229, 235, 237, 240, 242
    Bradoon 30, 33, 44, 83
    Buxton 204; Chifney 183
    Citation 11, 190; Curb 173, 237
    Double 75, 190; Fulmer snaffle 54
    Gag 44, 109, 111, 171; High Port vii, 43
    Leverage 31, 44, 191
    Liverpool 64, 70, 83, 116, 194
    Myler 43; Neue Schule 53, 105
    Pelham 82, 111
    Snaffle 47, 55, 134, 183, 218, 229
    Sprenger 43; Tom Thumb 54
    Waterford 111; Western 237
    Weymouth 43-4, 75, 98
biothane 38, 82, 85, 105, 124, 126, 142, 146, 147, 240
Black Bears Polo Team 169-171
Blair Castle 215, 218, 220, 112
blinkers vi, 30, 34, 36, 65, 70, 114, 116, 142, 183, 217, 229, 232
bosal 236, 237, 238
Bowman, George 87
breeching 40, 52, 60, 65, 67, 116, 118, 142, 144, 147, 159, 164, 204, 216, 220, 229, 242
Breen Equestrian 211-212
bridle
    bitless 30, 41, 54-5, 110-1, 214, 236-7, 242
    bitless, Dr Cook 55
    double 30, 32, 33, 98, 128, 132, 175, 177
    Liverpool 64, 70, 83, 194, 242
    Hackamore 242
bridleways 157, 158
Brind, Philip 69
British Horse Loggers 140-149
British Horse Society xii, 98, 108, 157, 158, 176, 240
British Schools of Racing 192
Brooke Hospital, Cairo 79
buckles 8, 29, 35, 41, 47, 52, 60, 65, 100, 114, 178, 201, 202, 204
Bunn, Douglas 208-211, 214

Byerley Turk 78, 185, 189
Cambiaso, Adolfo 166, 168
canal horses 65-71
Capel Manor, Enfield 14, 15, 30, 204
carbon fibre xii, 18, 28, 45, 100, 170, 207, 239, 240
carriage
    Bennington 85, 195
    driving 80-87, 193, 195, 199, 198, 205, 206, 225, 227, 240
    Fenix Carriages 17
    Foundation 83
    gun carriage 76, 202
    National Trust Carriage Collection 83
Cavalry 72-79
Cavendish, William, Duke of Newcastle 26, 28, 94-97
Cecil, Sir Henry 186, 187
CELT (Centre for Equine Learning and Therapy) 174, 176,195
Chanel, Coco 1, 24
Chariots of Fire Centre, The 199
Charles II 16, 42, 72, 75, 96, 97, 181
Chesters Estate 48-49
Cil Llwyn Farm 52-53
circus 88-93
Clwyd Special Riding Centre Ltd, Llanfynydd 193, 194-197
coach
    Royal Mail 81, 118
    stage coach 81, 87
    coach horses 81
    Diamond Jubilee 207
Coachmakers, Worshipful Company of 16-17
Coach Harness Makers, Worshipful Company of 16-17
Coakes (née Mould), Marion 210, 211
collars 36-38
    American 37, 67, 70, 16, 118, 143
    breast 36, 37, 38, 40, 86, 201
    English 35, 37, 38, 39, 65, 83, 114
    French 37
    Scandinavian 37, 38
    show 63
    zinc 37
Cordwainers, Worshipful Company of 11, 15, 16, 29, 34, 204
Crellin, Michael 231
cross-country 108-112
crupper 28, 35, 40, 142, 215, 220, 242
Cumbrian Heavy Horses 155, 156-157
Curriers, Worshipful Company of 4, 11, 15
Darley Arabian 185, 189
Dettori, Frankie 184, 199
Devil's Horsemen, The 225-227
Dolbadarn Film Horses 227
donkeys 57-60
Douglas Bay Horse Tram Company 229-232
Draught, Principles of 36, 38-40
dressage 94-101
    classical 92, 94-98
    modern 98-101
Driving 80-87
    competitive 84-86
    trials 85
Endurance 30, 45, 102-107, 119, 178, 182, 235, 240, 241

Epona 21, 61, 64
Epsom Derby 50
Eventing 9, 13, 98, 99, 108-112, 132, 139, 178, 192, 206
Exell, Boyd 84, 85, 86-87
Fairfax, Vanessa 29
Farriers, Worshipful Company of 11, 15, 17
Fenix Carriages 17
film horses 221-228
Fitzwilliam (Milton) Hunt 127, 133, 135-136
Flair System 110
Fortune Centre, The 199
Frankel 187
Gambia Horse & Donkey Trust 36
gait analysis 21, 29, 240
Gelli, livery stables 53-54
Gibsons of Newmarket 9, 46, 181, 182
Gidden, W & H 9
Gifford's Circus 90-93
girth 92, 96, 100, 105, 114, 130, 132, 134, 156, 169, 170, 183, 196, 212, 213, 215, 216, 220, 225
Godolphin Arabian 185, 189, 238
Grand National 53, 180, 184, 187, 188, 190, 208
Greatwood Centre, The 192, 198-199
hackamore 30 110, 212, 214, 242
hames 38, 39, 40, 64, 65, 70, 114, 116, 119, 142, 204, 229, 242
harness racing 122-126
harness
    Amish 116, 146, 147
    plough 40, 65, 142
    pole 40
    Red Morocco state harness 201, 202
    Scandinavian 37, 38, 40, 142
    shaft 40
    show 62, 64, 116, 118, 119, 120, 240
    trace 40, 142
Haute école 92, 94, 221, 222
Henderson, Nicky 189, 190-191, 192
Hermès xii, 2, 24, 25, 170
Hester, Carl 43, 96, 98-100, 110, 177 199, 240
Hickstead 9, 208-212
Hickstead Derby bank 210
Hickstead Derby 212
Hippotherapy 196
HM The Queen 72, 75, 76, 80, 130, 134, 182, 186
Hollesley Bay Colony Stud 118-121
hopples 122, 123, 124
horse brasses 35, 64, 118, 240
Horse Brass Society 118
horse logging 140-149
horseshoe 8, 17, 169
Household Cavalry Mounted Regiment 72-79
Hovingham Hall 97, 239
HRH Prince Charles, Prince of Wales 71, 140, 167, 192, 200, 201, 206
HRH Prince Philip, Duke of Edinburgh 80, 140, 167, 206
HRH Princess Anne, the Princess Royal 15, 16, 193, 194, 206
hunting 127-139
Injured Jockeys Fund 181
International Federation for Equestrian Sports (FEI) 107, 240
Isle of Man tram horses 229-232
Jockey Club, The 180, 181
Johnstone, Jim 142, 146, 147

INDEX

Joiner, Doug 140, 143
Jones, Dylan 227, 228
Jones, Glyn 52, 117
Jones, Pam 193
Keegan, Terry 40, 118
King, Mary 109, 110, 111
King's Troop Royal Horse Artillery 72-79
Lambourn 182, 188, 189, 190-191, 192
lifeboatix
Livery Guilds/Companies 11-17
Llewelyn, Colonel Harry 208, 209
long gears 140, 141, 142, 143, 144, 149
Long Riders 150-158
Lorinery 41-47
    National Collection of Lorinery 16, 42
    Worshipful Company of Loriners 15-16, 44
Manderston 50, 51,129
martingale viii, 31-32, 105, 134, 169, 171, 183
    running 105, 134
    standing 40, 169
Master Saddlers, Society of vi, vii, 9, 10, 13, 14, 15, 16, 24, 29, 30, 36, 47, 152, 211, 240
    Coppens, Catrien 204
    Creaton, Nick 26
    Davis, Terry 36, 37, 63
    Dempsey, Laura 22, 24
    Godden, Richard 13, 24
    McDonald, John 204
    Parkinson, Tiffany 182
    Roche, Frances 204
    Romain, Mark 13, 14
    Russell, Issi 29, 30, 31, 33, 100
McCoy, Sir Anthony 189
Mill Reef 185, 186
Milton Hall, the Saddle Room 132-135
mining 162-165
Mounted Police 159-161
Mull Pony Trekking Centre 54-56
museums
    Museum of the Horse, Tuxford 42, 235
    Museum of English Rural Life, Reading 121
    National Horse Racing Museum 181
    Walsall Leather Museum, Walsall viii, 26
    Weald and Downland Living Museum, Chichester 121
    Suffolk Punch Trust, Woodbridge 119
Musgrave of Belfast 134-135
Naprous, Gerard, Daniel and Camilla 225
National Horse Racing Museum 181
National Trust xi, 83, 84, 145
Native Americans 1, 235, 236, 238
Naylor-Leyland, Sir Philip 133
neck reining 41, 238
neck strap 32, 74
Newton, Thomas 8
nosebands
    cavesson 30, 33, 96, 134, 171
    drop 24, 33, 134, 171
    Flash 109
    Grakle 33, 171, 190
O'Reilly, Cuchullaine 150
Ortlieb saddlebags 152
Pacing 122-126
packhorse 65-71
Park House Stables 184, 186
Philips, Zara 109, 206
Philpot, Roger 22, 24
Piggott, Lester 184
pit ponies 162-165
Pittards plc 1, 5-6
Pliance System 19, 29
ploughing viii, 113-121, 142, 144, 157, 219

polo 10, 19, 32, 166-173, 178, 192
    snow polo 168
    rules of 171-172
polocrosse vi, 9, 166-173, 178
Pony Club v, vii, 16, 51, 53, 55, 110, 112, 166, 174-179, 240
Pony Express 103, 150, 234
Pressure mapping 19, 21, 29, 31, 37, 240
Princess Royal College of Animal Management and Saddlery, Enfield 14, 15, 30, 204
Queen Victoria 24, 25, 66, 130, 134, 173, 200, 201, 202, 205, 218
Queen's Life Guard 72, 79
racing
    Flat 180-192
    National Hunt 180-192
Rare Breeds Survival Trust 121
Rashid, Mark xii, 240
Reddaway, William 151, 153
Rehabilitation of Racehorses xii, 189, 192, 198-199
reining 41, 237, 238
reins 31
    bunny ear 195
    draw reins 169, 171, 173, 240
    Fantastic Elastic 100
    ladder 124
    rainbow 196
    romal 237
    split 237
Riding for the Disabled (RDA) 193-199, 206
Roberts, Monty xii, 43, 240
Rose, Annie 156
Royal Dragoons 79
Royal Mews xii, 15, 16, 17, 25, 35, 42, 200-207, 239
Royal Warrant 9, 24, 35, 46, 76, 85, 110, 134, 182, 196, 201
Roycroft, David 143
RSPCA ix, xii, 83, 89, 128, 163, 188
Saddlers, Worshipful Company of 12-15, 18, 24, 204, 206
saddle fitter 29, 240
saddle horn 22, 28, 173, 225, 234, 235, 236, 237
Saddlery Training Centre, Salisbury 13, 14, 24
saddle tree 1, 8, 10, 14, 15, 18, 19, 20, 21, 24, 41, 42, 45, 182, 234, 240, 242
saddles
    antique 25
    Army Universal Pattern 19, 28, 73, 77, 132
    Australian stock 55, 152, 172, 173
    cart vi, ix, 36, 63, 67, 114, 119, 132, 242
    Cossack 223
    deer 215
    dressage 19, 21, 225
    exercise 182
    holy 7
    hunting 1, 19, 22, 28, 46, 53, 130, 134
    jousting 223
    pack 28, 152, 226
    pad 27, 28, 45
    racing 46, 181, 182-183
    Roman 28
    side saddle vi, 1, 12, 13, 21-25, 45, 51, 52, 92, 93, 128, 129, 130, 200, 226
    Western 19, 26, 45, 221, 233-238
Saunders Stud 123, 124
Schaffler, Steffi 143
Schwarzenbach, Urs 168, 169
Scythian saddle 1, 27, 28
Sedgwick & Co, J & E 4-5
show jumping 208-214

Side Saddle Association 24, 25
Sister Mary Joy 174, 175, 176, 177, 199
Smart, Yasmine 88, 91, 92
Smythe, Pat 208. 209, 211
Society of Master Saddlers vi, vii, 9, 10, 13, 14, 15, 16, 24, 29, 30, 36, 47, 152, 211, 240
Sovereign Quarter Horses 237, 238
SPANA (Society for the Protection of Animals Abroad) 36, 79
Spanish Riding School 9, 95, 97
Speddyd Farm 51-52
St Moritz 169
stalking, deer 215-220
Stark, Ian 5, 98, 109, 111, 112, 177, 191
stirrup
    bars 10, 14, 19, 22, 23, 24, 45, 46, 47, 58, 183, 202
    leathers 1, 28, 45, 46, 96, 109, 128, 129, 134, 136, 170, 172, 182, 183, 211, 215, 236
stunt horses 221-228
Suffolk Horse Society 120, 123
Suffolk Punch Trust 119
swingle tree (see whipple tree) viii, 38, 39, 40, 63, 65, 67, 70, 71, 141, 142, 143, 144, 149, 145, 242
Synchronicity System 45
tanning 1-7, 236
Tattersalls 181
terret 41, 201
Tevis Cup 103
therapy with horses 93, 193-199
Thorne, Nicki 102-5, 107
Thoroughbred Rehabilitation Centre 192
Tindall (née Philips), Zara 109, 206
Tiverton Canal Co 69-71
traces 38, 39, 40, 65, 67, 70, 71, 83, 85, 114, 116, 118, 141, 142, 144, 185, 204, 229, 242
tram horses xii, 229-232
Trec 157
tree, saddle 1, 8, 10, 14, 15, 18, 19, 20, 21, 24, 41, 42, 45, 182, 234, 240, 242
trotting 122-126
Tschiffely, Aimé 150, 151
Ulvin's logging arch 140-145, 148
vaulting 89, 91, 92, 101, 175, 193, 195, 225
Wadworth Brewery 61-64
wagons xi, 40, 60, 62, 65, 80, 83, 91, 113, 115, 120,
Walsall viii, 4, 8-11, 15, 18, 26, 29, 30, 35, 46, 47, 100, 105, 110, 170, 195, 231, 232, 237, 239
Waterer, Jonathan & Fiona 114-117
Western riding centre 238
whip 8, 62, 82, 84, 87, 92, 124, 130, 194, 222
whipple tree (see swingle tree) viii, 38, 39, 40, 63, 65, 67, 70, 71, 141, 142, 143, 144, 149, 145, 242
Winsley Hall 49, 50
Winter, Fred 190, 208
World Horse Welfare xii, 36, 79, 194, 196, 206, 241
World War II xii, 1, 6, 8, 28, 50, 52, 57, 133, 137, 164, 191, 231
Wormwood Scrubs Pony Centre 174, 175, 176, 178, 199
Worshipful Companies
    Coachmakers 16-17
    Coach Harness Makers 16-17
    Cordwainers 11, 15, 16, 29, 34, 204
    Curriers 4, 11, 15
    Farriers 11, 15, 17
    Loriners 15-16, 44
    Saddlers 12-15, 18, 24, 204, 206
Wykeham pad 23